Daniel Hack Tuke

The Past and Present Provision for the Insane Poor in

Yorkshire

Daniel Hack Tuke

The Past and Present Provision for the Insane Poor in Yorkshire

ISBN/EAN: 9783337371289

Printed in Europe, USA, Canada, Australia, Japan

Cover: Foto ©berggeist007 / pixelio.de

More available books at **www.hansebooks.com**

THE PAST AND PRESENT PROVISION

FOR THE

INSANE POOR IN YORKSHIRE,

WITH SUGGESTIONS FOR

THE FUTURE PROVISION FOR THIS CLASS.

(Read at the Leeds Meeting of the British Medical Association.)

BY

D. HACK TUKE, M.D., F.R.C.P., LL.D.,

President of the Psychology Section.

London:

J. & A. CHURCHILL,

11, NEW BURLINGTON STREET.

1889.

DELIVERED AT THE OPENING OF

THE SECTION OF PSYCHOLOGY.

At the Annual Meeting of the British Medical Association, held in Leeds, August, 1889.

In taking the honourable post assigned me in this Section to-day, I must express the obligation under which I lie to those who have thus given me an opportunity and a powerful motive for dwelling on a subject congenial to my tastes and associations on the one hand, and germane in a high degree, as I venture to think, to the object of our meeting and to the particular county in which we assemble.

The subject I have chosen—the Past and Present Provision for the Insane Poor in Yorkshire—forcibly recalls to my mind the strong objection I felt to the name given to this Section when it was instituted. Surely it ought to have been Medical Psychology or Psychological Medicine, rather than "Psychology," pure and simple, a name which has the effect of making it seem out of place to mention bricks and mortar, and justifies the disgust which an eminent French alienist experienced when at our Glasgow meeting he had to listen to a debate on the best kind of cement for use in lunatic asylums.

Had I realized when I selected this subject for my Address the amount of time and labour it would involve, I should hardly have had the courage to attempt it. The mass of material it is necessary to study, the imperfection of many of the early returns of lunacy, the defective statistical tables in the Blue Books until a recent period, the errors which in the course of years have crept into the Asylum Reports, are enough to dishearten any historiographer of the insane poor who desires to compare their numbers and location at different periods. Why, only to mention one fatal omission in the Blue Books, there is no return of the admissions of the insane into workhouses, without which information we are totally unable to gauge the amount of occurring lunacy in regard to a large area of the pauper population.

I will, however, endeavour to make the best of the materials
I have been able to bring together. In a subsequent paper
I shall place before you a number of statistical tables, some
of the results of which I will mention in this address. I
would express to the medical superintendents of the Yorkshire asylums my heartiest thanks for the loan of their
Reports, and their kindness in supplying me with any information within their power which I required.

My review of the provision made for the insane poor in
Yorkshire must commence with a reference to the Lunatic
Hospital at York, generally known as the Old York Asylum.
From its foundation as a charity in 1777 it has been a mixed
institution, and received, in accordance with its rules, the
indigent poor whether paupers or not. It has also admitted
persons of limited means, and a certain proportion of opulent
patients for the purpose of assisting the funds. The Lunacy
Commissioners have repeatedly counselled the restriction to
private cases, and the establishment of a Borough Asylum
in York for pauper lunatics. We have no means of knowing
what the number of this class may have been when the
asylum was opened 112 years ago, but we may be very sure
that as there were only 15 patients altogether in this asylum
the year afterwards, the relief afforded must have been
entirely disproportionate to the need. The numbers very
gradually crept up till there were about 50 at the close of
the first decade ; and passing over twenty years more, we
find that there were 150 in 1809, but how many of these
were of the indigent class it is impossible to say. Now in
this year a pamphlet, entitled "Observations on the present
state of the York Asylum," was published, which reveals the
prevalent ignorance as to the actual amount of pauper lunacy
in this county, for there were those who at the time "actually
supposed that this institution was capable of receiving all
the needy and indigent applicants from the various parts of
the British dominions." This extraordinary belief it is the
purpose of the pamphlet to combat, and the writer observes :
"The futility of this will appear from the returns made to
the House of Commons preparatory to the Act of Mr. Wynn
in 1807. From this it may rather be inferred that with
respect to the lunatic parish-paupers alone, the asylum, even
on its present enlarged scale, would scarcely be found adequate
to their reception."

As the population of Yorkshire at that period (1809) was
about 950,000, the number of insane poor must have been

very considerable. Where, one asks, were they located? Some were detained in poorhouses, such as they were, but a large proportion of dements and idiots were with their friends, probably in some outhouse, attic, or cellar, or were wandering about, the butt of village ruffians, and were sometimes dangerous epileptics. Lastly, not a few lunatics were persons whose malady was unrecognized, and who were in gaols. Undoubtedly some were executed as criminals.

From the date of the opening of the York Asylum to 1814, the year when it was reformed and entirely reorganized, 2,635 patients were admitted, 2,133 were discharged, and 399 died, leaving 103 under care. How many of the admissions were of the pauper class cannot be discovered. In fact, Dr. Thurnam found it impossible to make any use of the statistics of this asylum during this period, in consequence of the imperfect state of the registers.

The average number resident appears to have been a little under 100. Since 1814 the institution has been a credit instead of a disgrace to those concerned in its management. It is a striking fact that the rate of mortality, which up to that year was at least 11 per cent. (calculated on the average number resident), was only about seven per cent. during the succeeding quarter of a century.*

I am able to state that in the following year there were 47 pauper patients from Yorkshire, 31 of whom were West Riding cases, five from the North, seven from the East, and four from the City of York. I mention these small items because they are very important, as showing the number of insane paupers provided for at this asylum in a year—1815—to which I must especially draw your attention, because it was signalized by the projection of the West Riding Asylum at Wakefield. Let me, however, conclude my notice of the York Asylum by the remark that its subsequent history has been one of great usefulness in accordance with its charitable design, that the old cells have long disappeared, and that it is now an attractive institution, containing 135 patients on January 1st last, 46 of whom were paupers. For many years it had, as you are aware, the advantage of having Dr. Needham as its medical superintendent.

To justify the erection of a county asylum, an attempt was made to ascertain the number of insane poor in the West Riding. There was reason to believe that there were at least 600 of this class, and it was supposed by Dr. (after-

* "Statistics of Insanity." By John Thurnam, M.D. 1845. App., p. 7.

wards Sir William) Ellis, the first superintendent at Wakefield, that there were 750 in Yorkshire altogether, without
proper accommodation for more than 200.   This accommodation was found, according to Dr. Ellis, in " a great many
private houses," as well as the York Asylum ; but I believe
that even he quite over-estimated the amount of this accommodation.   Thus he speaks of the York Asylum accommodating
100, but there were in it at that time only 47 pauper patients.
To us it sounds as incredible that there were only 750 pauper
lunatics in Yorkshire in 1815, as it then seemed incredible
to Yorkshiremen that there could be so many.   Taking his
estimate, the proportion to the population of the county
would be 1 in 1,300, whereas now it is 1 in 560.   It seems
strange to us nowadays to find Dr. Ellis laying down the rule
that from 100 to 120 patients are as many as ought to be in
any one house.   Where they are beyond that number, he
adds, the individual cases cease to excite the interest they
ought.   He further says : " It is absolutely necessary that
to manage such a house and such inhabitants the heads of
it ought to possess the most sovereign authority over all the
rest, and, consequently, to be accountable for everything.
The effects arising from the want of such an arrangement in
the Asylum at York, and from the introduction of it into the
Retreat, have surely been sufficiently exemplified " (page 17).
What would Ellis have thought had a prophet told him that
the asylum he superintended would, in 1889, contain nearly
1,400 patients, for the care and treatment of which one
unhappy superintendent is responsible?

The provision for the insane poor made at Wakefield was
a portion of the harvest gathered in this field from the seed
sown at York in 1792, when the Retreat was projected.   It
was its legitimate child.   That event had its constructive
and destructive aspects.   On the destructive side was the
revolt against the barbarous treatment of the patients in the
York Asylum, ending in its being purged of its abuses in
1814.   On the constructive side was a reformed hospital,
and then the desire to supply the needs of the insane poor
on a much larger scale.   Whatever share my own ancestors
may have had in this movement, they would have desired
that honour should be done to that bold and upright West
Riding magistrate, Godfrey Higgins, who not only aided
them in exposing the cruel neglect of the patients in the
York Asylum, but for years took an active part on the Committee of the Wakefield Asylum.   Having from boyhood

heard him spoken of in terms of the greatest respect and admiration, I rejoice to have this opportunity of reviving his memory in the Riding in which he dwelt and in the hearing of men who are so well aware of the contrast between the treatment of the indigent insane in 1815 and 1889. I hold in my hand an original letter written by Mr. Higgins to my father in April, 1815, which is very brief, but announces a fact of the greatest importance. It reads thus: " I write in great haste to inform you that it was the unanimous wish of the magistrates (at the Quarter Sessions) to accede to the proposal to build a place for our pauper lunatics, and we have proceeded as far as it was in our power according to law ; indeed, I believe a little further. There was but one opinion." The Wakefield Asylum was constructed to accommodate 150 patients on what is called the H plan, and was opened in November, 1818, seventy-one years ago.* I could not take a better illustration of the advance which has been made in asylum construction during this period than the architecture of this institution as originally constructed and that of recently-built county asylums, such as that erected at Menston.† It was, I make bold to say, from no want of thought or any lack of desire to do the very best thing for the care and comfort of the insane poor that allowed of such grave defects in the building, but those who planned it lived during the earliest stage of development in asylum architecture. It was necessary, I suppose, to follow the tedious but wholesome law of evolution rather than reach at one bound the highest form of art; and so it comes to pass that a building which in 1818 was justly regarded as a great step in advance would now be considered to be altogether below the standard of excellence. The comparison is not affected by the opinion held by not a few, that some of these modern asylums for paupers do not appear to be exactly suited for the particular class for which they are built.

The Wakefield Asylum, as is stated in an early annual Report, was judiciously built within one mile of the central

* " Practical Hints on the Construction and Economy of Pauper Lunatic Asylums; including Instructions to the Architects who offered Plans for the Wakefield Asylum and a Sketch of the most approved Design." By Samuel Tuke. 1815.

† It is greatly to be regretted that the Annual Reports of our Asylums carefully avoid giving the ground plan and elevation of the institution. This omission contrasts unfavourably with the excellent practice of the American Committees of Hospitals for the Insane. Is it unreasonable to hope that in future these plans may be provided in the Reports of British asylums ?

town of the West Riding, in a cheerful and healthy situation, combining the convenience of contiguity to a market town with the salubrity and quietude of the rural districts.

A very brief period elapsed before it was only too evident that the accommodation provided by the new asylum was altogether inadequate to the wants of the West Riding. It would be wearisome to enumerate the successive enlargements of the original building which were found to be necessary. Suffice it to say that its subsequent history is that which has become so familiar to us all, the record, namely, of the erection of building after building, here a little and there a little, to supply the unexpected demand made for admission of patients. In one of the Reports (1844) the superintendent, Dr. Corsellis, comments, like his predecessor, on the importance of not making asylums so large as to prevent minute observation in each case. He would have limited the number to 300, which he points out had already been much exceeded by the Wakefield Asylum, the average number of patients in that year having been about 400. In his Report two years later he refers to the apparent increase of insanity, but he does not hesitate to attribute it to "the beneficial operation of new laws by which attention was drawn to what was formerly regarded with culpable indifference, and a suitable provision for the insane poor enforced on parish officers." In that year (1846) the population of the asylum reached about 450. Ten years witnessed an alarming increase, the number of patients advancing to close upon 800; while in 1871, just before the opening of the second West Riding Asylum at Wadsley, the highest number in the history of the institution was reached, namely, 1,494. The number resident on January 1st of this year was 1,354. Mr. Cleaton's appointment to the post of medical superintendent in 1858 was the means of vastly improving the condition of the asylum, while his successors, Dr. (now Sir James Crichton) Browne, Dr. Herbert Major, and Dr. Bevan Lewis, have worthily sustained the reputation of this great institution. It would be interesting to dwell upon the good work that has been done at this asylum, administrative, clinical, and pathological; but I must rigidly limit myself to a somewhat dry record of events bearing on the real object of my address. I reluctantly, therefore, pass on to chronicle the erection of the South Yorkshire Asylum at Wadsley, near Sheffield, approvingly described by the Lunacy Commissioners as "a substantial structure of brick

and stone ; the architecture and elevation are of a plain but
varied and pleasing character, without unnecessary orna-
mentation." It was built to accommodate from 750 to 800
patients, and opened in the month of August, 1872. It
admitted from that time to the close of the year 321 patients,
and by 1874 there were 605 inmates. Ten years after its
opening—namely, in 1882—there were about 1,300 patients
in the establishment, and on January 1st, 1889, there were
1,616. Thus, again, has asylum history repeated itself, and
the necessity of the provision made at Wadsley for the
southern division of the West Riding has been more than
justified by the event. Of the first superintendent I will
only say that as the West Riding Asylum at Wakefield was
fortunate in having Dr. Ellis, so that at Wadsley was fortunate in securing the services of Dr. Mitchell.

As I am speaking of the asylums of the West Riding, I
will complete what I have to say about them and defer my
notice of the North and East Riding Asylums, which, in
point of chronological order, possess a prior claim. The
population of the West Riding has enormously increased,
in proof of which it is only necessary to state that while at
the census of 1801 it stood at about half a million (576,336),
it is estimated at the present time to be two and a half
millions (2,515,886). The consequent demand for additional
accommodation for pauper lunatics led to the erection of the
third West Riding Asylum at Menston, opened in October,
1888. The buildings now erected will accommodate 840
patients, those for chronic cases, designed, but not erected,
another 250 or 300 of each sex, making a total of some 1,440
patients. The health and success of the first medical superintendent, Dr. McDowall, in his arduous and responsible
position, we all, I am sure, most heartily desire.

In the construction of this asylum, we have undoubtedly
gratifying evidence of the praiseworthy desire of the magistrates to make provision for the pauper lunatics of the
district. It will be admitted that they have steered clear
of that false economy which, in erecting a second-rate
structure, involves in the course of a few years such frequent
and expensive repairs as necessitate in the end a still greater
outlay. I gladly recognize and pay my humble tribute to
the feeling which has thus actuated the county justices in
their plans of this large asylum. I hesitate to say a word,
more especially at the present crisis in the history of our
county asylums, which can be construed to imply anything

like blame ; I hope, however, I shall be forgiven if I venture
to say, bearing in mind the wisdom of avoiding even the
appearance of evil, that it would have been wise to avoid
anything in the way of costly embellishment calculated to
prejudice the mind of the ratepayer on entering the building.
Although he may not always discriminate fairly between
lavish expenditure and good, durable workmanship, I should
find it difficult to deny that, without the expenditure of so
large a sum of money, the structure might have been as
substantial, along with that just regard for cheerful appearance and even a certain amount of decoration which I am
sure none of us wish to ignore in institutions for a class of
persons for whom it is most desirable to avoid cheerless and
depressing surroundings.   When visiting this fine asylum I
could not help thinking that it may very possibly be utilized
some day, to a certain extent at least, as an asylum for the
middle and even the upper classes, some of whom at the
present time occupy much humbler quarters in private
asylums and registered hospitals.   It is not everywhere that

> " Each hospitable chimney smiles
>   A welcome from its painted tiles ;
>   The parlour walls, the chamber floors,
>   The stairways and the corridors,
>   The borders of the walls and walks
>   Are beautiful with fadeless flowers
>   Which never droop in winds or showers,
>   And never wither on the stalks."

To some of the poor labourers and peasants of our county it
is probable that a more homely—I had almost said a more
homeish—dwelling would have been quite as acceptable and
no less adapted to the great object in view, the comfort or
the cure of its inmates.

Those who planned the Menston Asylum, wisely adopted
the block or pavilion system, admitting as it does of complete
classification and the separation of patients labouring under
different states of insanity, as also the free circulation of air
between the various buildings.   There must, however, be
the means of passing readily between the central residence
of the chief medical officer and the wards of the sick and
the acute cases ; and this involves, if, as at Menston, the
blocks are at a considerable distance one from the other, an
extent of outside covered passages, which seem a too pro-
minent, and can hardly be regarded as an agreeable, feature

of the building. The extent to which complete separation from the administrative offices should be carried if the block system be adopted, is not a point on which anyone would like to speak dogmatically who knows what can be plausibly urged in favour of annectent gyri between all the psychic centres of the great brain of a county asylum; but I venture to hope that before the remaining blocks at Menston are built, full consideration will be given to the advantages of avoiding outside passages altogether, seeing that the patients residing in such blocks will be of the chronic and more or less incurable class. If at Kankakee, Illinois, which I visited with great pleasure a few years ago, this plan is carried out to an extreme, I think that we may learn a lesson from the experiment in avoiding the evils of our present system, and allowing of more structural distinction and a little more autonomy in the various buildings of our monster establishments. I trust also that when Menston is completed there will be a sensible difference in internal construction—and, therefore, the cost—between the blocks designed for the incurable, or rather the demented, class and those provided for the acute cases and the sick, which might be more on the hospital lines. Diversity in unity, differentiation in the form of the buildings and the proportion of single rooms, seem to me the great points to be borne in mind in the completion of this still unfinished asylum, it being also remembered that there is a limit to the expenditure on architectural designs, provided always that nothing be done in a penny-wise and pound-foolish spirit, or that will needlessly offend the eye.

This very imperfect sketch of the good work which has been done by the West Riding in providing three large county asylums must suffice, and I proceed to describe in a very few words what has been effected by the other Ridings of Yorkshire.

The asylum for the North and East Ridings, situated at Clifton, near York, was opened in April, 1847, and was constructed to accommodate about 150 patients. In its architecture it presented a marked contrast to the Wakefield Asylum, built about thirty years before.

The decision to provide this institution was arrived at by the magistrates of the North and East Ridings three years before it was completed. Some of the pauper patients in the York Asylum were removed to the new asylum, and the Committee of the former expressed a hope in their annual report that their institution "might ere long be relieved from the care of the greater part, if not the whole, of these —30 in number—who still remain in it."

The number admitted into the North and East Riding Asylum during the first year was 140. In the course of a decennium the numbers, as a matter of course, ran up, and amounted to 445, about 300 in excess of the original calculation. Another decennium passes, and we find close upon 500 under care, and, four years later, namely, in 1871, the number had reached 560. Then it was that the crowded condition of the building found relief by the opening of the separate asylum for the East Riding at Beverley. The consequence was that ten years after this event, the number of inmates in the North Riding Asylum was less instead of more, namely, 540; while on January 1st of the present year it was 622.

I should like to point out, to the credit of our county, that the decision of the magistrates to provide this asylum anticipated the compulsory enactments contained in Lord Ashley's Act, requiring every county in England to build (if it had not already done so) an asylum for the insane poor. The first superintendent, Mr. Samuel Hill, in his report of 1848, illustrated the wisdom and benevolence of the Ridings by the cases admitted when it was opened, which urgently required attention and treatment. One was that of a female removed from a private asylum, who, when under detention there, was described by the Lunacy Commissioners as "perfectly tranquil; her hands, however, were fastened by wrist-locks to a belt round her waist, to which a chain about nine feet in length was attached, and the end of the chain was padlocked to the floor near the wall." On the very day of her admission to the new county asylum, needlework was put into her hands, and she was from that time, with few exceptions, regularly employed. The only restraint resorted to was to have her hands tied behind her for an hour or so when disposed to commit violence. Although described by the proprietors of the asylum from which she was removed as "perfectly murderous," she became "as much under moral control as any other patient."

In this sketch of the North Riding Asylum at Clifton (York), the materials for which have been kindly placed at my disposal by Mr. Tregelles Hingston, I have incidentally mentioned that at Beverley. I must now add that the magistrates arrived at the conclusion to provide this institution for the East Riding in 1865. Six years elapsed, however, before it was completed, it being opened on October 25th, 1871. I have Dr. Macleod's authority for stating that it has proved a good, workable building. A hundred patients were at once admitted, and before the end of the year nearly all the patients in the North Riding Asylum who belonged to the Unions of the East Riding were transferred to Beverley, the total number admitted between October and December 31st being 197. In the course of ten years the number under care reached 277, and on January 1st of the present year there were in the house 291 patients.

The Hull Borough Asylum was originally a licensed house belonging to Dr. James Alderson and Mr. Richard Casson, surgeon. The date of opening is not certain, but there is a record of admissions which shows that patients were admitted as early as 1818, and it is probable that there were none previously.

Of its history for the next 20 years I have nothing to record. At the end of 1839, during which year 62 patients were admitted, there were 90 inmates. In the year 1842 the name of the institution appears as the Hull and East Riding Refuge, Mr. Richard Casson being the proprietor.

In July, 1849, the Refuge became the Hull Borough Asylum, having been purchased from the proprietor by the Hull borough magistrates. At that date there were 74 residents in the asylum. Ten years afterwards, namely, at the end of 1859, the number of patients was 85. Dr. Merson, to whom I am indebted for these particulars, and to whom the satisfactory state of the existing asylum is really due, accounts for the smaller number of cases in the institution after it became the Borough Asylum by the existence of an auxiliary institution for about 25 private and pauper patients, called Field House, at Anlaby, near Hull, which was under Mr. Casson's charge. It was closed in the year 1856.

The average number of patients in the Hull Borough Asylum from 1849 to 1868 was 103; ten years afterwards (1878) the number was 155.

The present asylum was opened with nearly 200 patients in December, 1883. It is situated six miles from Hull, at Willerby. The accommodation provides for 350, and the number at present in residence is 300.

In this survey of the asylums provided for the insane poor, it must not be forgotten that Yorkshire has every reason to be proud of the share it has taken in the provision for the education and training of idiots and imbeciles, although it does not possess an establishment for this purpose in the county itself. The Royal Albert Asylum at Lancaster may be regarded as largely Yorkshire in relation to its origin and the support which it has received, as in the instances of Sir Titus Salt, who gave 5,000 guineas, and the Rev. R. and Mrs. Brooke, of Selby, whose munificent donation amounted to £30,000.

During last year the contributions to the Maintenance Fund from Yorkshire amounted to £1,606, in the form of annual subscriptions and donations, being only about £300 less than the contributions received for the same account from Lancashire. It is to be hoped that a still larger amount will in future be raised in response to the appeal made in the last annual report. It should be remembered that, of 553 patients in that asylum, about 180 belong to Yorkshire. About one-fourth of the 120 pauper patients in the establishment are paid for by their unions in Yorkshire. Forty-two inmates were from Leeds. It is, however, thought by the authorities, including Dr. Shuttleworth, that Yorkshire, like Lancashire, ought to have its own pauper idiot institution, as there are in the union houses of the county numbers of young imbeciles who are wasting their lives for lack of training. In the last Report of the asylum, signed by Lord Winmarleigh, the opinion is expressed, "The time has arrived when separate provision for the care and training of pauper imbeciles should be made by the County Councils or some other local authorities, in the same way as obtains in the Metropolitan District under the Metropolitan Poor Act, 1867."

It may be stated that, according to the census of 1881, there were in Yorkshire 2,903 idiots and imbeciles, whether paupers or not, being at the rate of 1 in 997 of the population. The proportion in England and Wales was higher, namely, 1 in 794.

It should here be noted that a considerable number of pauper patients in Yorkshire were admitted into private asylums in addition to those under care in public institutions. Unfortunately there is nothing to boast of in regard to the way in which these patients were treated in these private houses. The numbers rapidly diminished after the opening of the North and East Riding Asylum. In 1847 there were 232 insane paupers in the Yorkshire private asylums. Five years afterwards there were only 69, while ten years later there was only one; since 1878 there have been none.

Of the provision for the insane poor made at different periods in this county in workhouses I must now say a few words. I shall not weary you as I have wearied myself in endeavouring to obtain definite information on this subject. I could hardly devise a more cruel punishment for my worst enemy than to have to wade through the Blue Books of the Poor Law and Local Government Boards. The imperfect and often contradictory statistics are bewildering. The Reports do, however, serve to show in what a lamentable condition the insane and idiots were when the Commission was appointed in 1832 to inquire into the administration and operation of the Poor Laws, while the subsequent Reports of the Poor Law Board show the increased care taken of pauper lunatics in workhouses. A debt of gratitude is due to a gentleman who from the earliest period laboriously worked at this subject, and still lives at the age of 90 to witness the complete change in the condition of English workhouses. I refer, of course, to Mr. (now Sir Edwin) Chadwick, who first served as one of the Commission, and afterwards became Secretary to the Board. I am sorry that I cannot claim him as a Yorkshireman, but he was very near being one, as he sprang from the adjoining county of Lancashire.

In 1836 Parliament ordered a return of the number of pauper lunatics and idiots in each county (the first of the kind). Now, the return for Yorkshire amounted to 1,046 lunatics and idiots, of whom 364 were in the County Asylum at Wakefield and the York Lunatic Asylum, while there were 88 in private asylums. The remainder, 594, were detained in workhouses or resided with their friends. Taking the population of the county at that period and the total number of pauper lunatics given in this important Parliamentary return, there was a ratio of at least 1 to 1,417.

These figures, we may well suppose, err on the side of omission, and, therefore, there must have been a large amount of lunacy outside asylums requiring better care and treatment than it received in workhouses. There have been some very bad ones in this county, on which the Lunacy Commissioners have reported most unfavourably; but I have no reason to believe they are other than in a satisfactory condition at the present time. Of the bad ones, in the past, that at Huddersfield may be taken as an illustration, even as lately as 1857. The Commissioners reported that it was "shamefully defective." One-third of the inmates were lunatic or idiotic, many of them were dirty in their habits and slept together, and "sometimes even in a perfect state of nudity. Those patients who were not disabled bodily were crowded together, and associated with most abandoned characters, secluded in the deadhouse, and were so circumstanced in other respects as to end in the most deplorable state of degradation." It is needless to say that all this has been changed, but it is very necessary that we should remember what was the condition of lunatics in workhouses, in many instances, until comparatively recent times.

As an illustration of a workhouse which has, on the contrary, presented a good record, I may take that at York. I do not refer to it as being a model house, or absolutely the best in the county, but because it is known to me as an excellently-conducted place; and I have at hand particulars kindly supplied to me by Mr. S. W. North, for many years the medical officer in attendance. There are about 120 patients, and during the last fifteen years the annual admissions have averaged 26. Some were admitted more than thirty years ago. The form of mental disease is one for which a workhouse is obviously well suited. Fifty-five of the above number were imbeciles, 41 dements, 19 chronic maniacs, three idiots, and one only a melancholiac. I will not add more to these figures than the statement that, of 250 discharges during the same period of fifteen years, 82 were sent to their friends, 37 to an asylum, and 131 died. I do not think that any of you would maintain that this class of patients ought to have been cared for in an expensively-built asylum. I have inquired particularly into the cost. It seems to be impossible to obtain accurate information, the reason being that the insane inmates are part of the general family; no person is exclusively devoted to them; their diet is drawn from the common stock and is cooked with the rest; this diet, at the same time, is not all alike—it can only be estimated in the lump, not individually. A comparison with the cost of maintenance in an asylum is, therefore, most difficult. The master of the workhouse says that a charge of 5s. per head weekly on the whole number of lunatics would cover everything, except building, with a margin to spare. Mr. North, in a valuable paper which appeared in the "Journal of Mental Science," in July, 1882, estimates that, while the the charge for buildings would in asylums be from £10 to £15 per head annually, it would not be more than from £3 to £4 in workhouses. Mr. North calculates that the saving to the ratepayers of the York Union by properly utilizing the workhouse for insane paupers is not less than £1,000 a year. He states that "it is the custom in this Union, and the desire of the Guardians, to send to the asylum every case which, in the opinion of their medical officer, needs the special care of an asylum; and, so far as I know, no single case has ever been kept in the workhouse to the prejudice of the patient." What Mr. North very properly contends for is, "that there should be some careful and skilled discretion exercised, in the first instance, as to who should and who should not be sent to an asylum."

But the time at my disposal will not allow me to do more than add to what I have said of the provision for the insane outside asylums, that the number at the present time (January 1st, 1889) in detention in Yorkshire workhouses is 1,316, or 22·26 per cent. of the grand total of 5,879* pauper lunatics in Yorkshire; while the number resident with their friends or elsewhere is 330, or 5·58 per cent. The majority (4,265) are cared for in asylums, amounting to 72·16 per cent.; while in England and Wales the percentage is lower, namely, 69·01 per cent. The statistical tables I have prepared show that the recovery rate in the Yorkshire asylums has been very respectable, namely, 39·69 per cent. of the admissions (including transfers); while the corresponding percentage in the county and borough asylums of England and Wales during the decade ending 1888 was 35·60. Excluding transfers (for as long a period as they are specified), the recovery rate in Yorkshire asylums has been 41·90; while in England and Wales it has been lower, namely, 40·13.† Then as to the mortality rate, this in the county and borough asylums of Yorkshire has been 12·09 from their opening to the present time, calculated on the average number resident.

---

* The total number belonging to Yorkshire is 5,911, but all are not resident in the county.

† Table IV. shows the net recoveries in those asylums in which the proper distinction is made between Cases and Persons. Were the Hull Borough Asylum taken separately, it would be found that during about 40 years, 1,834 persons were admitted, omitting transfers. Of these 690, or 37·62 per cent. were discharged recovered. Now of these 293 were re-admitted relapsed, leaving 397 not relapsed. Of the relapsed patients, 188 were discharged recovered, the net result yielding the lower percentage of 31·89 per cent.

In the similar asylums in England and Wales it has been (during the last decade) 10 per cent. (9·99). But it must be borne in mind that the period during which the death-rate in Yorkshire extends dates from 1818, and that if we had the corresponding period for England and Wales instead of during the last ten years, it would in all probability be as high as in this county. You may remember that the eminent statistician of whom Yorkshire can boast as a native, the late Dr. Thurnam, of the Retreat, York, gives a table showing that the annual mortality of patients in English pauper asylums from 1812 to 1844 was 13·88, and observes that, "considering the frequently distressed and too often depraved condition of the pauper population of the manufacturing districts of Yorkshire and Lancashire, we need not, perhaps, be surprised that the mortality of the asylums of Lancaster and of the West Riding of York has materially exceeded that which has been estimated as the maximum mortality."*

And now, looking back over the period our survey has embraced to the date when that philanthropic and rational movement commenced, which led, among other things, to the erection of asylums for the lunatic poor in this county, we naturally ask, Has the object for which they were established been fulfilled ? I have no hesitation in answering the question in the affirmative. This unhappy class of persons have been rescued from neglect and inhuman treatment, and are now well housed and fed, kindly treated, and placed under medical care in admirably-managed institutions. In fact, when criticism is made, it is to the effect that philanthropy has gone to an extreme, and has involved an unfair and disproportionate outlay upon the insane, as compared with the sane, sick pauper.

But while we rejoice in this great and beneficent result, I cannot conceal from myself the fact that the pioneers in this work entertained hopes which, if they could be with us to-day, they would sorrowfully admit had never been realized. I am sure they would say that they fully expected a large proportion of the insane under the reformed treatment to be permanently cured of their malady, and never for an instant supposed that of those who should recover, a frightful proportion would relapse, and that, seventy years after the opening of the first pauper asylum in Yorkshire, there would be five large county and one borough asylum in addition to

* "Statistics of Insanity," p. 138.

that at York. I venture to say that had it been possible for them to have foreseen this, their astonishment and dismay would have been great indeed; and, were they present at this moment, it would seriously temper the satisfaction they might justly experience in witnessing the success of their humane labours. They would find us engaged in endeavouring to explain the apparent increase instead of decrease of insanity, as shown in the tables I have prepared; and, although there are explanations which suffice to show that such an inference is hasty and superficial, our ghostly visitants would find, if they interviewed the Superintendent of the Wakefield Asylum, that his knowledge of the masses from which his patients are derived, leads him to form a very unfavourable opinion of the mental status of many who are outside our asylums—that, in fact, there is a large borderland class who are practically insane, although able to adapt themselves fairly well to their environment, usually relatives of patients already inmates of the asylum, and often far more insane at times than the latter. Surely, to the prevention no less than the cure of insanity must our endeavours be bent in the future.

In concluding this Address, which in accordance with the custom of the Association is supposed to be undebateable, I would observe that it would be deprived of much of any value it may possess if it did not lead up to a practical consideration of the lessons which the history of the past should teach us in regard to the future provision for the insane poor in this, or, indeed, in any other county in England. This question I reserve for the paper I am about to read, which, I trust, will elicit the opinions of those whom I address, and to whom I shall appeal for suggestions.

It now only remains to thank you for the patience with which you have listened to my history of a movement which deserves an abler, though not a more appreciative historian than myself, and to express my ardent hope that the work of this Section will be second to none of the others in its interest and utility.

(Statistical Tables.—See page 31.)

ON THE PROVISION FOR THE INSANE POOR IN THE FUTURE.

Should the future provision for the insane poor in Yorkshire, and not in this county only, but in other counties of England, be conducted on the same lines as the past provision has been?

This paper aims at being something more than historical, and faces the practical questions which naturally arise from the survey which I have just made.

There are several questions we should try to answer.

I. Can the boarding-out of pauper patients with strangers be introduced into Yorkshire on a sufficiently large scale to sensibly relieve asylums from the pressure which, judging from the past, will be their experience in the future?

II. Is it possible to relieve the pressure on asylums by paying a more liberal sum for the maintenance of harmless and incurable patients in their own homes?

III. To what extent ought workhouses to be utilized? Has this utilization been carried as far as it suitably can be in providing accommodation for incurable lunatics?

IV. Closely connected with this question is the consideration of the operation of the Capitation Grant.

V. Lastly, would it have been wiser to build two asylums instead of one, so as to provide a comparatively small establishment for the recent and acute cases, and another of a less expensive character for the demented, and, as part of the same question, should the smaller institution be on the same estate and under the same management as the other?

Unless we are prepared to maintain that the provision made for the insane and idiotic is as perfect in its character as the wit of man can devise, that there is no reasonable probability of preventing the enlargement of existing institutions by developing the home or cottage care of this class, or providing more lunatic wards in workhouses, or that if more asylums have to be built they cannot be planned more wisely than in the past, I say, unless we are justified in taking this optimistic position, there is an ample field for practical advice from those who are constantly engaged in attending to the needs of the indigent insane. Even if we decide that nothing better can be done in the future than has been done in the past, there will be the satisfaction of having obtained this testimony from those who are competent to judge.

I. Should anyone present chance to have read my article on the " Boarding-Out System in Scotland " in the " Journal of Mental Science " of January last, he will allow that I am no enemy to the system, but that, on the contrary, I am likely to be hopeful rather than despondent as to the effect of its application to the congested lunatic districts of Yorkshire. And when one reads what Dr. Rutherford says in his last Report, that by adopting this remedy the number of pauper patients in the asylum belonging to the district in which it is located, so far from increasing, has actually diminished during the period between 1869 and 1888, I say, when we see such a happy result as this, one cannot but ardently desire to see a similar course pursued in this county, provided only that it be practicable, · beneficial to the patients, and (what is too often overlooked) not harmful to the families in which they reside.

Before, however, recommending the system of boarding-out pauper lunatics as one means of preventing or lessening the accumulation of this class in public asylums, it is surely wise to ascertain what conclusions have been already arrived at in this matter by those on the spot competent to judge, and who are only too anxious that it should succeed. Now, Dr. Mitchell, the late Superintendent of the Wadsley Asylum, and Dr. Major, of the Wakefield Asylum, seriously contemplated carrying out the Scotch system some years ago. They came, however, to the decided conclusion that it was not practicable to adopt it in Yorkshire, mainly on account of the character and density of the population they had to deal with. Dr. Bevan Lewis is unfortunately driven to precisely the same conclusion, and holds that the objections to boarding-out in Yorkshire are legion, that is, using the term " boarding-out " in the sense in which it is generally employed, of placing patients with strangers for profit.*

II. Although, however, boarding-out the insane poor on the Scotch system appears to be impracticable, or, at least, injudicious in Yorkshire, a good deal has been done in sending from the asylums suitable cases home to their friends, which although one form of boarding-out, is better understood as the " trial-out system." This meets with Dr. Bevan Lewis's approval. Again, in the annual report of the Wadsley Asylum for 1883,

---

* I have vainly endeavoured to discover how many of this class are scattered about in Yorkshire, but the number is stated to be extremely small. I have obtained returns from Scarborough, Barnsley, Doncaster, Halifax, Pontefract, Sheffield, Holbeck, Wakefield, Wharfdale, Wortley, and N. Brierley, arranged under three heads :—In Lodgings and Boarded-out, 1. With Relatives, 175. Alone or in Almshouses, 9. Total, 185.

it is stated that 46 patients had been discharged to the care of their friends, having improved or become so feeble that they could with safety and convenience be entrusted to the care of relatives. In the reports for 1886 and 1887, Dr. Mitchell states that the decision of the Committee to grant to patients discharged on trial a weekly sum equivalent to the cost of their maintenance in the asylum, has been attended with satisfactory results, Dr. Mitchell observing that " as only a small proportion of the cases thus dealt with have been sent back, the result of the efforts made to avert, by such means, the pressure on the accommodation of the asylum and to meet the requirements of fresh cases, may be said to have been fairly successful."

Dr. Merson, of the Hull Borough Asylum, whose experience is quite in accordance with that of Dr. Mitchell, informs me that many of the harmless cases are, in his opinion, much better with their friends, when the latter can be induced to take them.   In Hull, he finds that the friends, as a rule, are willing, and even anxious, to take charge of such cases without remuneration.   During the ten years of his residence there, he has admitted 722 pauper patients.   Of these, 100 who were deemed incurable and suitable for home-life were given up to the care of friends, with the exception of about ten sent to the workhouse.   Sixty-seven have not been returned to the asylum at all.   Thirty-three were sent back to it, but most of them after a considerable time.   Of these, 11 were again entrusted to the care of their friends, so that 78 of the 100 so treated are still outside the asylum.   There are no lunatics boarded out with strangers in Hull.

If the friends are able to take back these chronic cases without remuneration, well and good, but I think that, as a rule, it will be found necessary to give it.

I am glad to see that Dr. Chapman, in his last report of the Hereford Asylum, has dwelt strongly on the point of granting more liberal allowance to the relatives of patients kept at home, and recommends the grant of 5s., 6s., or 7s. a week, whereas at present it rarely exceeds half-a-crown.   He advises that for all pauper lunatics " the County Council should have power to determine whether a patient should be placed in an asylum, the workhouse, or with their friends or others, with power to order the allowance in the latter instance, and to arrange for adequate inspection."   This is just one of the points in which we may be of use in advising the new authorities in regard to their proceedings in the care of the insane poor.

You are aware that Dr. Duckworth Williams, the late Superintendent of the Sussex County Asylum, staved off the evil day of having to erect additional buildings by making provision for a considerable number of quiet and harmless cases outside the asylum. I have obtained from him the exact result of his praiseworthy endeavours in this direction, and, seeing that what he has done has proved so successful, it is probable that the same plan may be pursued, to some extent at least, in this county. I say "to some extent," because conditions may exist in Yorkshire which are not so favourable to the success of the experiment as those which exist in Sussex.

Dr. Williams was appointed Superintendent of the Hayward's Heath Asylum in 1869, when the beds were being rapidly filled up by an annual excess of 25 admissions over discharges and deaths. As the population of Sussex was steadily rising, a still greater increase was to be expected. Well, during the 18 years Dr. Williams was Superintendent, he, by encouraging, or rather pushing, the discharge of chronic and harmless cases to the care of their friends, reduced the average yearly increase to 10 patients, and by that means saved the county the expense of building another asylum.

From time to time cases were picked out which were considered fit for trial, and having ascertained that the relatives were both willing and able to receive the patient, Dr. Williams recommended him for discharge in the usual way. He was able to report after a trial of $2\frac{1}{2}$ years that the anticipations entertained of the feasibility of this plan had been most encouragingly fulfilled. Fifty cases were discharged to the care of their friends, with the result that 23 remained with them, only six returned to the asylum, two died, one had to go to the workhouse, while 10 started again in life to resume their former avocations. (Result unknown in eight cases.) During the whole period of 18 years ending Jan. 1st, 1888, 600 chronic lunatics, or 15 per cent. of the admissions, were discharged either to the care of friends or to the workhouse. How many of these went to the latter, and how many to their friends, I cannot ascertain. During the six years ending Jan 1st., 1888, however, 132 of 277 discharges were sent to the workhouse and the remaining 145 were placed under the care of relatives.

In the evidence which Dr. Williams gave in 1877 before a Select Committee of the House of Commons, he urged an amendment of the present laws by which the boarding-out of pauper lunatics with relatives and others might be facilitated. I think his proposals were very wise, namely, that the county asylum should be the head-quarters of all the pauper lunacy in the county; that all lunatics should be on the books of the institution ; that all chronic and harmless cases boarded with relatives or sent to workhouses should be retained on the asylum register. It was also proposed that they should be visited weekly by the Union Medical Officer, and quarterly by a medical man on the asylum staff, who should have the power to at once send away on his sole authority any case which he might consider should be readmitted into the asylum. Further, that the Union Medical Officer should at once report to the Medical Superintendent any change he might notice in the patients he visited, and that the Committee of the asylum should have power to order an allowance to be made to the person with whom a lunatic is boarded.

We have now a complex Lunacy Act in which these wise proposals have been to a certain extent recognized. Section 40 enacts that when the Asylum Committee is satisfied, on application from the relative or friend of a patient confined in an asylum, that he will properly care for him and is approved by the Guardians,* the patient may be delivered over accordingly, and the authority liable for the maintenance of the lunatic shall pay to the person to whom the lunatic is delivered such allowance, not exceeding the expenses incurred in the asylum, as such authority may on the recommendation of the Committee of Visitors think proper. It is also provided that the patient should be visited once in every three months by a medical officer of the district of the Union in which the lunatic is resident, and such medical officer shall within three days after each visit report the result thereof to the Committee of Visitors to the asylum. It will be admitted that this is a very important section. Dr. Williams informs me that if his proposals had been adopted when he was Superintendent, he could have easily doubled the number of outdoor lunatic paupers. While I do not believe that what is true of Sussex would to anything like the same degree be true of Yorkshire, I am heartily glad that facilities and inducements will soon be in operation, along with some, at least, of the safeguards which are absolutely essential to the suitable carrying out of the system of out-door relief to chronic pauper lunatics.†

* If the proposed residence is outside the limits of the Union to which the lunatic is chargeable, the approval of a justice having jurisdiction in the place where the relative or friend resides must also be obtained.

† Sub-section 5 of this Section enacts that so long as an allowance is paid, the lunatic shall not be deemed a pauper lunatic in an asylum for the purposes of the Lunatic Act, 1853, Sect. 66.

As to the proportion of cases which are likely to be cared for either by their friends or others, I think that at the very outside it will be 10 per cent., more probably eight. At present it is about five per cent. in Yorkshire, and seven or eight per cent. in England and Wales.

III. There can be no doubt that very great relief indeed has been afforded to the Yorkshire asylums by the judicious use of the union workhouse. The extent to which workhouses can be properly utilized or even extended as substitutes for county asylums must always depend upon a careful selection being made of the chronic cases proposed to be sent there. Dr. Mitchell, in a letter recently addressed to me, speaking of his experience at the Wadsley Asylum, says that they were able to transfer far more patients to their respective workhouses than to the care of their relatives, and that he had the less hesitation in pressing this course as he knew that the imbecile wards of the West Riding Workhouses were, on the whole, very well managed, and he goes so far as to say that he had reason to believe that the patients were happier and more contented than they were in the asylum, as they could visit or could be visited by their friends much more frequently than when confined in the latter. In one of his annual Reports (1885), Dr. Mitchell expresses his regret that " the policy of providing for suitable cases in workhouse-wards which has the approval of the Commissioners in Lunacy, and which has been repeatedly advocated as not detrimental to the class of patients concerned, is not more generally followed. The immediate effect of refusal to act upon it on the part of any union may relieve local rates, but the course can hardly be right and just towards other unions, which adopt a different and, it is contended, a wiser view. One policy has the tendency to avert—the other to render constantly necessary—the call for costly asylum extension and new institutions out of proportion to the number of cases which require, or can be benefited by, the more expensive accommodation."

Dr. Bevan Lewis has kindly favoured me with his experience on this vitally practical question, and arrives at the conclusion that "there is no remedy for the relief of our choked up asylums beyond a far greater development of the lunacy wards of our union workhouses, better managed and officered than they now are." This is a very important observation, and when we speak of workhouses we ought to make it clear that we mean workhouses prepared for the admission and proper care of lunatics.

The Commissioners in Lunacy have on several occasions brought the unnecessary removal of chronic and harmless cases from workhouses to asylums under the notice of the Local Government Board, and have obtained from the superintendents a return of patients who might be taken care of in workhouses :—

1. In special lunatic wards.
2. In workhouse infirmaries with paid nurses.
3. In ordinary workhouse wards.

The returns thus obtained convinced the Commissioners in Lunacy that a very considerable proportion of the present asylum population might be adequately and more economically provided for in workhouses. It is gratifying to find that their inspection of workhouses satisfied them that on the whole " the treatment of the insane and imbecile inmates is fairly good, and that Boards of Guardians will frequently adopt reasonable suggestions for their improvement. In many workhouses, indeed, the services of these inmates are most valuable and save the cost of much paid labour " (39th Report, page 114).

Having for many years taken a special interest in the insane in workhouses, I may say that my experience is in accordance with these observations.

As to the percentage of cases which may properly find shelter in workhouses, it may be put at 25 per cent. at least. At present it stands in Yorkshire at 22·7.

IV. This brings me to the operation of the much-debated Capitation Grant of 4s. a week to the Guardians in aid of the maintenance of each pauper lunatic maintained in county and borough asylums. Unquestionably the original intention of the Legislature in allowing this grant was praiseworthy, having for its object the encouragement of Guardians to send recent and curable cases to asylums instead of retaining them in workhouses from motives of economy. Its operation, however, has extended far beyond this desirable result, and the feeling and experience of the Superintendents of the Yorkshire asylums have been, and are, strongly averse to this unfortunate grant.

I could bring an overpowering amount of testimony in support of this conclusion. As I am aware that some are still unconvinced, I must be allowed to quote the observation of the present Medical Superintendent of the Wakefield Asylum in a letter received from him a few months ago : " The Capitation Grant," he writes, " has been most demoralizing in Yorkshire, and my experience of the co-operation of union and asylum work has been far from cheering."

I learn from a recent letter from Dr. Mitchell that about eight or ten years ago the Halifax Guardians, influenced by the fact that this grant would come into operation, and being in want of room, actually appropriated the imbecile ward for general purposes, and sent Wadsley the whole of the imbeciles—some 140—of a class "wholly suitable for workhouse treatment, and who had been very well treated there." As the patients were near their former homes and relatives, they complained bitterly of being removed so far away from their old surroundings. These patients, Dr. Mitchell says, are still at Wadsley.

The remedy for the mischievous working of this grant lies, it seems to me, in extending its operation (if retained at all) to workhouses and cases of out-door relief. Other modes of re-apportioning the grant have been proposed, but I think that this re-adjustment would be the simplest, and would go far in effectually removing the temptation to transfer chronic cases from workhouses to asylums. In consequence of the recent Local Government Act, the Capitation Grant is now paid by the County Councils out of the funds they receive from the Probate Duty and Licenses which were formerly paid direct to the Imperial Treasury, but I have the best authority for stating that the County Councils have no authority for making any change in the mode of distribution of the grant. For this the sanction of Parliament will be required.

V. Passing on now to the last question, namely, whether it would have been wise to build two moderately sized asylums instead of one large one—the acute cases being located in one institution and the dements in the other—I have partly anticipated the reply in my Address when speaking of the Menston Asylum. I have no doubt that county lunatic asylums, wherever erected, should be more differentiated than was the case in former days. There should be the means of treating acute cases in a separate hospital block, one in the construction of which no reasonable expense ought to be spared, or there should be a hospital at some distance from the asylum on the lines laid down by Dr. Newington in his recent Presidential Address. For the chronic class of patients other buildings, constructed much more cheaply, will doubtless suffice. The number of patients must, in any case, be large, but it is hoped that the evils resulting from enormous masses of the insane being collected together in one building will be avoided in future more than it has been in the past, although I can by no means go so far as to assert with some that our present asylums are manufactories of dementia. Nor am I prepared to endorse the outspoken remark of the Medical Superintendent of one of our largest asylums, who said to me the other day : " The talk about separate asylums for the curable is all humbug. The advantage of our monster asylums is this—that a patient is left to himself ; whereas, if you'd tinkered him with drugs, you would have made a chronic maniac of him !" While fully approving, however, of the separate treatment of acute cases in a hospital on the same grounds as the other blocks or else-where, I confess that I am not so sanguine as to the curative results as many are. In considering the amount of provision which ought to be made for curable and recent cases of insanity, I must say at once that at the very root of this question lies a fact which we are slow to recognize, and, when recognized, are unwilling to admit, but one which, it appears to me, it is idle to ignore, namely, that "recent" and "curable" cases are sadly far from being convertible terms, that the curability of insanity has been greatly exaggerated, and that if the liability to relapse is honestly taken into account, we shall be able to understand the disappointment felt in regard to the results of the conscientious and praiseworthy county care which has been extended to the insane poor, and we shall be guided to sound practical conclusions as to the extent of asylum accommodation which we ought to place at the disposal of the curable insane. It surely behoves us, before we speak scornfully of the existing county asylum system, or indulge in optimistic expectations and prophecies as to the wonderful cures which will result from separate hospitals for recent and acute cases, to realize these facts. A passage in one of Dr. Major's annual Reports of the Wakefield Asylum enforces the painful truth for which I am now contending. He says : "I am constrained to express my opinion that in the large majority of the unfavourable cases admitted, the unhopeful prospect has been due, not to want of recourse to early treatment, but, so to speak, to inherent unfavourableness determined from the very outset of the mental symptoms. *Experience soon shows how numerous are the cases admitted into any asylum, in which the insanity is given (and so far as can be ascertained correctly given) as being of recent date, and yet in which but little or no hope of recovery can be entertained.*" (Report, 1878, p. 16).*

* I would also mention in this connection some important evidence given at the Select Committee of the House of Commons on Lunatics in 1859 by the Chairman of the Hanwell Asylum Committee (Sir Alexander Spearman), when he was asked : " Is not the effect of keeping patients in an asylum, instead of trans-

Dr. Merson has recently afforded evidence of the increased ratio of cases of general paralysis to the total number of admissions, and he is not alone in his experience. Whether the sum of occurring insanity is greater than formerly or not, there seems to be overwhelming evidence that this hopeless form of mental disorder has increased in frequency in Yorkshire and Lancashire. In his Report of the Hull Asylum, 1883, Dr. Merson says : " To show the hopeless character of many of the admissions I need only mention that more than 26·5 per cent. of the persons admitted were suffering from general paralysis." Again, in this year's annual Report, he observes that, of 824 cases admitted, 309 were suffering from forms of mental disease which rendered a cure from the first impossible. Of course these evidences of the hopeless character of a large number of cases of insanity admitted into our county asylums reveal very disagreeable facts, but unless they are candidly admitted and widely stated, the popular, and, I must think, medical illusion is perpetuated, that if only every case of mental disease were immediately placed under treatment, recovery would follow in a very large proportion of instances, some promising us even 80 per cent. The exposure of this fascinating fallacy, at once so pleasant to indulge in, and the cause of such bitter ' disappointment, does not, however, by any means disprove what has been very properly insisted upon, and by no one more strongly than myself, that prompt and judicious treatment in the early stage of mental disease will prove successful in some cases, which if they remained at home and received no special medical treatment would go from bad to worse, and lapse into a condition of hopeless dementia.

What I have said only affects the amount of the provision which ought to be made for curable cases, many holding that

ferring them to the workhouse, to keep out from your asylum curable cases to which you might do much good ? " The reply was in the negative, and he proceeded to state that the Commissioners in Lunacy had communicated to the Committee the fact that there were no less than 76 curable pauper cases in private asylums in Middlesex, and had expressed a wish that immediate arrangements should be made for receiving these curable cases into the Hanwell Asylum. The witness then stated that they undertook to remove into the asylum every case that was really one likely to be benefited by the removal, and by being placed under more careful management. The Asylum Medical Officers visited every one of those cases, and they reported upon them to the Committee. What was the result ? Why, to show that out of those 76 patients not more than half were cases in which there was the slightest possible chance of recovery. Four females were transferred to Hanwell, and in regard to these the Chairman's evidence was as follows :—" When I inquired yesterday it appeared that one had died and three were incurable." (Report of Select Committee, April 11, 1859, p. 251.)

the extent of this should be very large, and I holding that it need be only very small.

Annexes for the chronic class of cases have met with the general approval of Superintendents—the amount of land, the staff of attendants, and the outlay on the building being less than in curative establishments. They have been promoted by the Lunacy Commissioners from time to time. In one of their Reports they maintain that " The best mode of making provision for pauper lunatics for which asylums have no accommodation is to erect inexpensive buildings adapted for the residence of idiotic, chronic, and harmless patients in direct connection with, or at a convenient distance from, the existing institutions. These auxiliary asylums would be intermediate between union workhouses and the principal curative asylums. The cost of building need not in general much exceed one-half of that incurred in the erection of ordinary asylums, and the establishment of officers and attendants would be on a small and more economical scale than those required in the principal asylums " (1887). I have the high authority of Mr. Cleaton for stating that the Board has seen no reason to alter the opinion thus expressed in regard to one of the most effectual remedies for the evil of which we are so painfully aware, and are so anxious to mitigate—accumulation. (See Note, p.   .)

I would here record, emphatically, my admiration of the county asylums as a whole, for the success of which we are indebted to the Committees of Visitors as well as the Medical Superintendents. Well indeed will it be if their successors in authority equal them in their efficiency. As Dr. Needham has justly said : " The generally admirable condition of asylums has been largely due to the direct or indirect action of the Magistrates, who have been well served by men whom they have trusted." If the liberality we admire has sometimes been marked by lavish prodigality, I do not wish the ratepayers or County Councils to suppose that cheapness is necessarily the one thing to secure, and that anything is good enough for the insane poor. The late Mr. Gaskell, the Commissioner, for whom all who knew him entertained so deep a respect, and whose name is now permanently associated with an examination for honours in psychological medicine—Mr. Gaskell, more than 40 years ago, thus wisely expressed himself in regard to the danger of extravagance in the erection of pauper asylums :—" It will not, I trust, be imagined that I have any wish to advocate a stinted accommodation for the insane poor of the county (Lancashire). On the contrary, my ardent desire is to see a liberal amount of that accommodation provided ; but, at the same time, I submit that it should be judicious in its character and suited to the conditions of the parties needing it." Golden words, these !

Of the proportion of pauper lunatics for whom provision ought to be made in asylums I should place the minimum at 65 per cent., in that case apportioning 25 per cent. to workhouses, and 10 per cent. to outdoor relief, which is probably too high ; while I think the maximum proportion of asylum cases may be estimated at about 70 per cent. In Yorkshire it has reached 71 per cent.

In conclusion, I would summarize my conclusions as follows :—

1. That the resources offered by the system of boarding-out patients with strangers are, as regards this county, of a very restricted character, and, therefore, any sanguine hopes excited by what has been effected in Scotland by this plan will not be fulfilled. The practice of paying something to the friends of pauper patients towards their maintenance to a more liberal extent than is at present the case should be encouraged. That after all has been done that can properly be done in the way of out-door relief, the great mass of pauper lunacy remains a fearful tax on asylum accommodation.

2. That workhouses, under proper conditions, including separate lunatic wards and effectual supervision, should be used to the greatest possible extent for that hopeless and chronic class of cases, which experience has shown may be cared for with sufficient regard to their comfort and interests.

3. That the Capitation Grant should, if retained, be readjusted, so as to avoid offering a temptation to guardians to send chronic cases to asylums.

4. That after provision has been made in workhouses and in private dwellings the great majority of the insane poor, probably 65 per cent., must be provided for in county and borough asylums.

5. That either distinct blocks should be prepared as hospitals for presumably curable cases, or a separate hospital for this class at some distance from the asylum.

NOTE.—Experience has shown that even in Annexes, it is desirable to have a few curable cases. The interest of assistant medical officers is damped by attending to incurables only.

STATISTICAL TABLES

SHOWING THE PAST AND PRESENT PROVISION FOR THE INSANE
POOR IN YORKSHIRE, AND THEIR BEARING ON THE FUTURE.

TABLE I.—Summary of Pauper Lunatics in Yorkshire, Jan. 1, 1889.

Opened.	Asylum.	Number.
1777	York Lunatic Asylum	46
1818	Wakefield (W.R.)	1350
1847	York (N.R.)	573
1849*	Hull Borough	276
1871	Beverley (E.R.)	259
1872	Wadsley (W.R.)	1562
1888	Menston (W.R.)	167
	In Workhouses	1316
	With Friends and Elsewhere	330
	Total	5879†

* Year in which the asylum was purchased by the Borough. The new asylum was opened in 1883.

† The number in the Report of the Commissioners in Lunacy 1889 (Table IX., p. 32), derived from the returns made by Clerks of the Guardians of Unions and Parishes of England and Wales, is 5,911, including cases paid for by the county, on account of the parish being undetermined. There are no pauper patients in the Private Asylums of Yorkshire.

TABLE II.—Showing the Admissions, Discharges, and Deaths from the opening of the West Riding Asylums (Wakefield, Wadsley, and Menston), the North Riding, East Riding, and Hull Borough Asylums, to the 1st January, 1889.

	M.	F.	Total.	M.	F.	Total.
Cases Admitted				18,272	18,605	36,877
Cases Discharged—						
Recovered	6,498	8,137	14,635 (39·69 p.c.)			
Relieved	1,689	1,800	3,489			
Not Improved	1,059	1,012	2,071*			
Died	7,010	5,330	12,340			
Total Cases discharged and died since opening of the Asylums				16,256	16,279	32,535
Remaining 1st January, 1889				2,016	2,326	†4,342

* Includes Transfers to other Asylums. In the Wakefield Asylum Annual Reports, the "not improved" cases are included under the "relieved" up to the year 1850, and Transfers are not given until 1873.
† This total includes a few private and out county patients.

TABLE III.—Showing the total Admissions into the County and Borough Asylums of Yorkshire, distinguishing the Asylums from their opening to the 31st December, 1888, with the total number of Recoveries, and the percentage thereof, calculated on the Admissions, with and without Transfers.

Asylum.	Period.	Admissions.		Recoveries.		Percentage of Recoveries on Admissions.	
		Including Transfers. (1818-88)	Excluding Transfers. (1847-88)	1818-88.	1847-88.	Including Transfers.	Excluding Transfers.
West Riding Asylum (Wakefield)	1818 to 1883	20,294	—	8,585	—	42·30	—
Do.	1873 to 1888	—	7,613	—	3,227	—	42·39
West Riding Asylum (Wadsley)	1872 to 1888	7,430	6,411	2,671	2,671	35·95	41·66
West Riding Asylum (Menston)	Oct. 8, 1883, to Dec. 31, 1888	167	—	—	—	—	—
North Riding Asylum (Clifton near York)	1847 to 1888	5,296	4,786	2,096	2,096	39·58	43·79
East Riding Asylum (Beverley)	1871 to 1888	1,382	1,063	407	407	29·45	38·28
Hull Borough Asylum	1849 to 1888	2,308	2,266	876	876	37·96	38·61
Totals		36,877	22,139	14,635	9,277	39·69	41·90
County and Borough Asylums in England and Wales	1879 to 1888 ...	...	...	...	...	...	40·13
	Do. ...	...	...	...	...	35·60	...

TABLE IV.—Showing the Recoveries of *Persons* calculated on the *Persons* Admitted into the Hull Borough, East Riding and West Riding (Wadsley) Asylums, from their opening to January 1st, 1889.*

History of Recoveries of Persons.	M.	F.	Total.	The same, excluding Transfers.		
				M.	F.	Total.
Persons Admitted†	4,653	5,006	9,659	4,018	4,298	8,316
Of whom were Discharged Recovered during the same period, being 35·14 per cent. of Persons Admitted ...	1,431	1,953	3,384	1,418	1,943 (40·12)	3,360
Of these Recovered Persons there were Re-admitted Relapsed	297	459	756	295	459	754
Leaving Recovered Persons who have not Relapsed	1,134	1,494	2,628	1,123	1,484	2,607
Relapsed Persons Discharged Recovered	167	242	409	167	242	409
Net Recovered Persons, being 31·44 per cent. of Persons Admitted	1,301	1,736	3,037	1,290	1,726 (36·27)	3,016

P.S.—The materials for this Table cannot be obtained from the Annual Reports of the Wakefield, North Riding, and Menston Asylums.

* 1. Hull Asylum during 39½ years (1849-88).

2. East Riding Asylum, 17 years (1871-88).

3. West Riding Asylum (Wadsley), 16 years and 127 days (1872-88).

† The number of *Cases* admitted into these Asylums during the same period amounted to 11,120, and the Recoveries of *Cases* to 3,954, or 35·55 per cent., as against 31·44 per cent. when the calculation is made on *Persons*, as in the Table.

TABLE V.—Showing the Admissions, Discharges, and Deaths in the Yorkshire County and Borough Asylums, with the Mean Annual Mortality and Proportion of Recoveries per cent. since the opening of the Asylums to December 31st, 1888.

| Year (Dec. 31). | Admitted. | Discharged. | | | Died. | Remaining 31st December in each year. | Average Numbers Resident. | Percentage of Recoveries on Admissions (including Transfers). | Percentage of Deaths on Average Number Resident. |
		Recovered.	Relieved.	Not Improved.					
*1818 to 1867	13,556	5,465	1,002	345	4,900	1,844	721	40·31	13·87
1868	710	315	28	24	180	2,007	1,897	44·36	9·48
1869	710	298	24	15	247	2,133	1,989	41·97	12·46
1870	646	315	18	9	244	2,193	2,164	48·76	11·27
1871	855	298	84	139	263	2,264	2,392	34·85	10·99
1872	1,119	308	59	293	247	2,476	2,505	27·52	9·85
1873	968	338	49	109	321	2,627	2,563	34·91	12·52
1874	1,102	436	49	121	321	2,802	2,722	39·56	11·79
1875	1,162	471	52	86	368	2,987	2,882	40·53	12·76
1876	1,156	539	96	94	334	3,080	3,068	46 62	10·88
1877	1,120	464	137	60	369	3,170	3,121	41·42	11·72
1878	1,174	464	209	75	347	3,249	3,231	39·52	10·73
1879	1,206	533	149	85	370	3,318	3,316	44·19	11·10
1880	1,219	470	192	37	355	3,483	3,427	38·55	13.27
1881	1,238	501	175	32	396	3,617	3,601	40·46	10·99
1882	1,227	495	173	108	370	3,698	3,662	40·34	10·10
1883	1,237	486	183	39	395	3,832	3,778	39·28	10·45
1884	1,335	490	215	47	449	3,966	3,939	36·70	11·39
1885	1,132	482	139	48	484	3,945	3,941	42·50	12·28
1886	1,281	505	150	39	444	4,088	4,010	39·42	11·07
1887	1,282	492	129	69	487	4,193	4,131	38·37	11·37
1888	1,442	470	177	197	449	4,342	4,222	32·59	10·64
	36,877	14,635	3,489	2,071	12,340		1,455	39·69	12·09

* As there were no Discharges or Deaths during the short period the West Riding Asylum was opened in 1818, that year has been omitted from the calculations; the number of years up to 1887 has been reckoned as 49, and the number of years up to 1888 (inclusive) being reckoned as 70.

TABLE VI.—Showing the Mortality Rate of the County and Borough Asylums of Yorkshire from their opening to 1887 (inclusive), in decennial periods, calculated on the average number resident.

I. WEST RIDING ASYLUM, WAKEFIELD.

Period.	Deaths (Average Annual Number).	Average Annual Number Resident.	Deaths per cent.
*1819-28	30·9	192·8	16·02
1829-38	52·9	292·7	18·07
1839-48	51·0	421·5	12 00
1849-58	121·4	711·2	17·06
1859-68	155·7	1069·7	14·55
1869-78	154·2	1428·4	10·79
1879-87 (9 years)	146·4	1405·5	11·56
1819-87 (69 years)........	103·2	780	13·20

II. NORTH RIDING ASYLUM, NEAR YORK.

Period.	Deaths (Average Annual Number).	Average Annual Number Resident.	Deaths per cent.
1848-57	21·6	264·7	8·16
1858-67	46·4	479·4	9·67
1868-77	49·0	488·7	10·03
1878-87	59·9	557·1	10·75
1848-87	40·0	447·4	9·60

III. HULL BOROUGH ASYLUM.

Period.	Deaths (Average Annual Number).	Average Annual Number Resident.	Deaths per cent.
†1849-70			
1871-80	26·1	145·9	17·88
1881-87 (7 years)	29·8	210·1	14·18
1871-87 (17 years)...............	27·64	172·3	16·01

IV. WEST RIDING ASYLUM, WADSLEY.

Period.	Deaths (Average Annual Number).	Average Annual Number Resident.	Deaths per cent.
‡1873-82	105·3	897·9	11·80
1883-87 (5 years)	154·4	1438·6	10·37
1873-87 (15 years)...............	121·7	1095·5	11·11

V. EAST RIDING ASYLUM, BEVERLEY.

Period.	Deaths (Average Annual Number).	Average Annual Number Resident.	Deaths per cent.
1872-81	27·30	220·48	12·37
1882-87 (6 years)	29·00	279·24	10·39
1872-87 (16 years)............. ..	27·93	256·25	10·90
England and Wales 1878-87 (County and Boro' Asylums).	...	...	9·99

* The Asylum was opened in 1818, but not long enough to comprise any deaths.
† The Mortality Returns for 1849-70 are doubtfully correct, and therefore have been omitted.
‡ The year of opening (1872) comprising only four months is excluded.

TABLE VII.—Showing the Recovery Rate in the County and Borough Asylums of Yorkshire, from their opening to 1887 inclusive, in decennial periods, calculated on the total Admissions (including Transfers).*

I. WEST RIDING ASYLUM, WAKEFIELD.

Period.	Admissions.	Recoveries.	Percentages.
†1819-28	1,124	508	45·10
1829-38	1,430	607	42·44
1839-48	1,474	563	38·19
1849-58	2,991	1,285	42·96
1859-68	3,919	1,591	40·59
1869-78	4,842	2,239	46·24
1879-87 (9 years)	4,030	1,641	40·71
1819-87	19,810	8,434	42·57

II. NORTH RIDING ASYLUM, NEAR YORK.

Period.	Admissions.	Recoveries.	Percentages.
†1848-57	895	272	30·39
1858-67	1,244	521	41·88
1868-77	1,423	577	40·54
1878-87	1,453	639	43·97
1848-87	5,015	2,029	40·46

III. HULL BOROUGH ASYLUM.

Period.	Admissions.	Recoveries.	Percentages.
1849-70	969	395	40·8
1871-80	651	275	42·44
1881-87 (7 years)	579	172	29·69‡
1849-87	2,199	842	38·29

IV. WEST RIDING ASYLUM, WADSLEY.

Period.	Admissions.	Recoveries.	Percentages.
†1873-82	4,083	1,436	35·17
1883-87 (5 years)	2,525	1,014	40·15
1873-87	1,608	2,450	37·08

V. EAST RIDING ASYLUM, BEVERLEY.

Period.	Admissions.	Recoveries.	Percentages.
†1872-81	708	244	34·46
1882-87 (6 years)	408	145	35·53
1872-87	1,116	389	34·85

* In comparing Decennial periods it would obviously be unfair to include the first year in which an asylum was opened. This year has, therefore, been omitted in this Table, with the exception of the Hull Asylum, which was re-christened in 1849, but was not a new asylum.

† As the Recovery Rates in this Table are exclusive of the year of opening, while those given in Table III are inclusive of that year, there is a discrepancy to that extent between them. This Table, which gives the most favourable results, is fairer to the Asylums than the other.

‡ The low Recovery rate is due to a large number of Transfers from the East Riding Asylum, and the unfavourable character of many of the Admissions.

TABLE VIII.—Showing the Admissions of Pauper Lunatics into the County and Borough Asylums in Yorkshire (including Re-Admissions and Transfers) during the years 1849-1868 inclusive. ·

Years.	Total Admissions.	Years.	Total Admissions.
1849	397	1859	508
1850	346	1860	510
1851	441	1861	574
1852	399	1862	531
1853	449	1863	528
1854	436	1864	590
1855	369	1865	531
1856	423	1866	562
1857	525	1867	639
1858	501	1868	709

TABLE IX.—Showing First Admissions of Pauper Lunatics into the County and Borough Asylums in Yorkshire during the Decades 1868-77, 1878-87; also the Annual Average during each decade.

Years.	Ratio of First Admissions to Pop. (per 10,000).	FIRST ADMISSIONS.			
		Including Transfers.		Excluding Transfers.	
1868	2·60	590		2·55	583
1869	2·56	595		2·53	588
1870	2·29	542		2·27	537
*1871	2·28	550	Annual Average = 740	1·42	342
†1872	3·47	868		2·78	696
1873	2·67	665		2·16	537
1874	3·31	836		2·80	709
1875	3·58	926		3·19	825
1876	3·52	942		3·18	851
1877	3·28	892		3 05	832
1878	3·23	894		2·43	773
1879	3·38	952		3·09	869
1880	3·31	947		3·02	865
1881	3·43	998	Annual Average = 958	3·34	971
1882	3·20	948		3·13	927
1883	3·24	974		3·11	935
‡1884	3·31	1020		3·06	935
1885	2·83	881		2·74	853
1886	3·15	966		2·86	911
1887	3·14	1009		3·07	984

* Opening of the East Riding Asylum.
† Opening of the Wadsley Asylum.
‡ Immediately after the opening of the new Hull Borough Asylum.

TABLE X.—Showing the Ratio of Pauper Lunatics to the Population and to the Number of Paupers in various Counties, including Yorkshire, January 1st, 1888.

COUNTY.	Ratio of Pauper Lunatics to Population.	Ratio of Paupers to Population (per cent).	Ratio of Pauper Lunatics to Paupers (per cent).
Middlesex	1 in 261	2·70	14·21
Wilts	1 in 271	4·32	8·42
Gloucester	1 in 284	3·77	9·34
Kent	1 in 378	3·20	8·27
Norfolk	1 in 319	4·53	6·92
Southampton	1 in 323	3·65	8·49
Somerset	1 in 338	4·21	7·02
Warwick	1 in 338	2·40	12·30
Devon	1 in 339	4·31	7·02
Surrey	1 in 346	2·55	11·36
Sussex	1 in 403	3·62	6·84
Essex	1 in 428	3·46	6·74
Notts	1 in 430	2·50	9·28
Lancashire	1 in 456	1·98	11·07
Cheshire	1 in 474	2·37	8·88
Stafford	1 in 487	3·52	5·84
Yorkshire	1 in 560	2·32	7·54
Durham	1 in 668	2·34	6·38
England & Wales	1 in 384	2·90	8·92

TABLE XI.—Showing the Number and Distribution of Pauper Lunatics and Idiots in Yorkshire in 1837 in pursuance of an Address to the Crown July 5th, 1836. (First Return made by the Poor Law Commissioners.)

RIDING.	In Asylums.	In Licensed Houses.	In Work-houses and Out-door.	Totals.
East Riding......................	23	71	92	186
North Riding	36	14	126	176
West Riding.....................	305*	3	376	684
Totals	364	88	594	1,046 { 1 in 1,417 Pop.
England and Wales (at same date)......................	—	—	—	13,667 { 1 in 1,000 Pop.

* At that time there were in the Wakefield Asylum about 300 patients.
NOTE.—In Middlesex there was in the same year 1 to 696; in Lancashire 1 to 1,960; in Wilts 1 to 750; and in Surrey (lowest proportion in England) 1 to 1,965.

(Corresponding Return, January 1st, 1889.)

RIDING.	In Asylums.	In Licensed Houses.	In Work-houses and Out-door.	Totals.
East Riding.....................	574	—	207	781
North Riding	552	—	142	694
West Riding	3,139	—	1,297	4,436
Totals...................	4,265	—	1,646	5,911 1 in 560 †

† In 1888, 1 in 570. At the same date there were in the East Riding 1 in 475, in the North Riding 1 in 555, and in the West Riding 1 in 573.

TABLE XII.—Showing the number of Pauper Lunatics maintained in County Asylums, Registered Hospitals, Private Asylums, Workhouses of Yorkshire, and those residing with Relatives or elsewhere, Jan. 1st, 1861, 1871, 1881, 1888, and 1889. Also the number per cent. under these headings.

Years.	Pauper Lunatics.				Of each 100 Pauper Lunatics the number maintained.			
	In Asylums.	In Work-houses.	With Relatives, etc.	Total.	In Asylums.	In Work-houses.	With Relatives, etc.	Total.
1861	1571	697	379	2647	59·40	26·30	14·30	100·0
1871	2191	1064	407	3662	59·83	29·05	11·12	100·0
1881	3381	1274	302	4957	68·20	25·70	6·10	100·0
1888	4159	1297	325	5781	71·94	22·44	5·62	100·0
1889	4265	1316	330	5911*	72·16	22·26	5·58	100·0
England & Wales. 1889	53193	17509†	5930	75632	69·01	23·15	7·84	100·0

* See note, Table I.
† Includes the Metropolitan District Asylums. If these were deducted and added to the "Asylums" the percentages would be as follows :—In Asylums, 76·27 ; in Workhouses, 15·89.

TABLE XIII.*—Showing Date of Opening, Number of Patients, Expenditure, etc., in Yorkshire County and Borough Asylums, to Jan. 1, 1887.

YORKSHIRE.	Date of Opening.	Population of District, 1881.	Amount of Land.	Amount of Land Rented.	Total Amount of Land.	Cost of Land to December 31st, 1886.	Expense of Building to December 31, 1886.	Total.
			A. R. P.	A. R. P.	A. R. P.	£	£	£
†North Riding, near York......	April 7, 1847	315,756	158 0 0	—	158 0 0	22,542	109,423	131,965
West Riding, Wakefield	Nov., 1818	2,175,314	49 2 22	78 1 24	128 0 6	11,397	345,627	357,025
West Riding, Wadsley	Aug. 27, 1872		194 3 28	—	194 3 28	30,253	337,151	367,404
‡East Riding, Beverley	Oct. 25, 1871	111,690	113 3 19	—	113 3 19	5,798	57,825	63,623
Hull Borough (New Asylum)	Dec. 13, 1883	154,240	74 2 0	—	74 2 0	12,770	79,249	92,019
		2,757,000	590 3 29	78 1 24	669 1 13	82,760	929,275	1,012,035

* Extracted from the latest Parliamentary Return (1887). † Excluding Borough of Scarborough. ‡ Excluding Hull and York Boroughs.

TABLE XIII. (*Continued.*)

YORKSHIRE.	Accommodation, December 31, 1886			Number of Patients in Asylum, January 1st, 1887. Belonging to the District			Belonging to other Districts			Cost of Maintenance per head during 1886. Annual (£ s. d.)	Weekly (£ s. d.)	Officers	Salaries £	Allowances
	M.	F.	T.	M.	F.	T.	M.	F.	T.					
North Riding, Near York...	309	321	630	236	247	483	52	54	106	21 6 10	0 9 4	Supt.	800	Furnished House, Man-servant, Coals, Gas, Washing, Stabling, etc., £287.
												Asst. Med. Off.	170	Rations, Attendance, Furnished Apartments, Coals, Lighting and Washing, £126.
West Riding, Wakefield ...	695	703	1,398	693	704	1,397	1	—	1	22 8 10	0 8 7	Supt.	600	Board, Furnished House, three Servants, £420.
												1st A. M. O.	200	Board, Furnished Rooms, Attendance, £90.
												2nd A. M. O.	150	Ditto, ditto.
												3rd A. M. O.	130	Ditto, ditto.
West Riding, Wadsley	708	844	1,552	660	789	1,449	23	71	94	20 18 2	0 8 0	Supt.	800	Residence, Board, etc., £200.
												Senior Asst.	250	Ditto, ditto, £75.
												2nd Asst.	180	Ditto, ditto.
												3rd Asst.	120	Ditto, ditto.
												4th Asst.	120	Ditto, ditto.
East Riding Beverley	150	150	300	109	122	231	44	17	61	22 3 0	0 8 6	Supt.	500	Furnished House, Horse Keep, Washing, Coals, Gas, Garden Produce, £160.
												Asst. M. O.	100	Board, Lodgings, Washing, and Attendance.
Hull Borough	130	180	366	128	124	252	—	2	2	27 9 9	0 10 7	Supt.	400	Furnished House, Board, Washing, Coal, Gas, use of Garden, and Keep of Horse, £350.
												A. M. O.	120	Board, Lodging, Washing and Attendance, £70.
	2,012	2,193	4,240	1,826	1,986	3,812	120	144	264	117 6 7	2 5 0			

TABLE XIV.—Showing the Total Expenditure; Average Weekly Cost of Maintenance, Medicine, Clothing, and Care of Patients; and Weekly Charge for Patients in the County and Borough Asylums of Yorkshire during the Year 1888.

Ridings and Borough.	Total Expenditure.		Average Weekly Cost.										
	On Maintenance Account.	On Building and Repairs Account.	Provisions (including Malt Liquor in Ordinary Diet).	Clothing.	Salaries and Wages.	Necessaries: (*e.g.*) Fuel, Light, and Washing.	Surgery and Dispensary.	Wine, Spirits, and Porter.	Charged to Maintenance Account. Furniture and Bedding.	Charged to Maintenance Account. Garden and Farm.	Charged to Maintenance Account. Miscellaneous.	Less Monies received for Articles, Goods, and Produce Sold (exclusive of those consumed in the Asylum	Total Average Weekly Cost per Head.
	£	£	s. d.	s. d.	s. d.	s. d.	s. d.	s. d.	s. d.	s. d.	s. d.	s. d.	s. d.
York, N. Riding ...	15,691	4,845	3 7	— 7¼	2 6¼	— 8⅞	— 0½	— 1⅛	— 6½	1 4¼	— 3⅞	— 7¼	9 2
,, W. Riding (Wakefield)	32,223	5,237	3 4½	— 9¼	2 3½	— 8½	— 1⅛	— 0⅞	— 5	— 9⅜	— 2⅝	— 2⅞	8 5⅞
York, W. Riding (Wadsley)	34,371	9,577	3 1½	— 9½	2 2⅝	— 8¾	— 0¼	— 0½	— 5¾	— 5	— 1⅝	— 1½	7 10⅛
York, E. Riding......	6,253	696	2 3½	— 8½	2 5⅝	— 11⅞	— 0¾	— 0⅓	— 5⅞	1 7½	— 2⅞	— 5	8 6
Hull Borough Asylum.........	7,839	2,235	3 5½	1 0¾	2 10⅛	1 3⅝	— 0⅞	— 0⅜	— 4⅜	1 5	— 5⅛	— 5	10 7
	96,377	22,590											

TABLE XIV.—(*Continued.*)

RIDINGS AND BOROUGH.	Weekly Charge for Paupers from Counties or Boroughs to which Asylum belongs.	Weekly Charge for Paupers from other Counties or Boroughs.	WEEKLY CHARGE.		
			Weekly Charge for Private Patients.	Fund to which the Excess beyond the ordinary Weekly Charge is carried.	Fund to which the Payments for Private Patients are carried.
	s. d.	*s d. s. d.*	*s. d. s. d.*		
York, N. Riding	9 4	14 7 to 15 2	14 0 to 31 6	Additions	Additions, and Maintenance
„ W. Riding (Wakefield)	8 0	14 0	14 0	Maintenance	
„ „ (Wadsley)	8 0	14 0 to 12 6	20 0 to 14 0	ditto	Maintenance
„ E. Riding	8 9	15 0	13 0 to 30 0	Capital	Excess to Capital
Hull Borough Asylum	10 6	14 0	14 0 to 30 0	Maintenance	Maintenance

NOTES ON THE TABLES.

(With Additions).

As a mass of statistics involves the danger of "not seeing the wood for the trees," I have separated the less important statistical tables from the others, and would here briefly note the most salient results obtained in regard to the pauper lunacy of Yorkshire.

We have to deal at the present time with (in round numbers) 6,000 insane and idiotic poor, being one in 560 of the whole population of the county, and a ratio of pauper lunatics to paupers of 7·5 per cent. Now this, as will be seen by referring to Table X., shows that Yorkshire is much more favourably circumstanced than most other counties; in fact, out of 18 having an insane population of 1,000 and upwards, it has, with one exception, the lowest ratio.

As compared with England and Wales, Yorkshire stands out very creditably also, for in the former the ratio is 1 in 384. Table XVIII. exhibits the same fact, inasmuch as the ratio in the one is 17·85 per 10,000, and in the other as high as 26·06.

That the apparent increase in the number of the insane in proportion to the population has been enormous since the first official returns is indisputable, being from 1 in 1,417 in 1836 to 1 in 560 in 1889; but in the first place it is certain that the former estimate would be much below the mark; and, secondly, there is the inevitable accumulation which must occur, even if there is no actual increase in occurring insanity, so long as the annual recoveries and deaths fall below the admissions. It would not be fair, however, to assume that this necessary accumulation accounts for the whole of the increase which our tables exhibit. The question of increase of insanity (in excess of that of the general population) would be solved if we could ascertain with certainty the ratio of *occurring* insanity to the population at an early and recent period of the history of the county. But, unfortunately, it is absolutely impossible to obtain for a lengthened period this primarily important information. It is only since 1868 that returns have been made of the annual number of first admissions into asylums. These are taken as roughly—very roughly—representing the number of persons who have annually become insane since that year. Unfortunately, it does not include that large mass of patients who have become insane but have been detained in workhouses or are boarded out. If, disregarding these sources of fallacy, we take the first admissions (*minus* the transfers), we find that the ratio to the population (per 10,000) was in 1868, 2·5, and in 1887, 3, not a very large increase. But this only goes back 20 years, and therefore affords very imperfect data for any sound conclusion.

Again, it is obvious that if the mortality of the insane in asylums was greater formerly than at the present time, we should have one reason for the apparent increase of insanity. Table V. shows that it was greater, although not to the extent which might have been expected. The difference would, no doubt, be vastly greater if we were comparing the mortality-rate in good and bad asylums, whereas we are in this instance comparing good asylums in the past with good asylums of the present day. The result would also be very different were we comparing the old workhouse mortality with that in asylums or modern workhouses. Tables VI. and XV. show the mortality in Yorkshire asylums at different periods, and also in those of England and Wales during the last ten years.

Further, in regard to recoveries, if the proportion has decreased we shall have an additional cause of the accumulation of the insane in asylums. This has certainly been the case, although the difference is not great. Whether this difference arises from the more sanguine view taken by our predecessors as to the recovery of patients, or whether the apparent increase of general paralysis accounts for the circumstance, the result is the same in its bearing on the accumulation of the inmates of asylums.

But by far the most important factor in the large accession of patients to asylums is the transference which has taken place to them from workhouses and from their homes. This is well brought out by Tables XI. and XII. Thus, in 1836, 43·22 per cent. were in public or private asylums, and 56·78 per cent. in workhouses or were outdoor patients; while in January, 1889, 72·16 per cent. were in asylums, and 27·84 in workhouses, or were receiving out-door relief.

The following four tables bear on the question of the increase of insanity, and speak for themselves. The first and second show how far the discharges and deaths have fallen short of the admissions. It will be seen that in the third and fourth of these tables, the figures comprise the *total existing* numbers of insane poor in Yorkshire, wherever detained, at different periods, and form a valuable contribution to the statistics of pauper lunacy in this county. The succeeding table (Table XIX.) shows that between 1847 and 1888 there were close upon 3,000 re-admissions into the Yorkshire asylums. Upwards of 1,800 admissions were transfers. These figures demonstrate the fallacy of all calculations made upon the gross admissions into asylums.

Since the tables were completed, I have obtained the following items in reference to the Menston Asylum:—Amount of land, 318 acres; total average weekly cost per head, 8s. 11¼d.; salary of Superintendent, £400 with board, rooms, and attendance; salary of 1st and 2nd A.M.O.'s, £120, with same extras. Cost of building up to present time, not communicated.

It may here be stated that the other County Asylums in Yorkshire have cost in land and buildings £238 per bed. (See Table XIII.)

TABLE XV.—Showing the percentage of Pauper Lunatics in County and Borough Asylums in Yorkshire, Discharged, Recovered, Relieved, Unimproved, or Transferred, calculated on the Admissions, and the percentage of Deaths calculated on the mean number resident between 1818-67, and during the decades 1868-77 and 1878-87.

	DISCHARGES.		DEATHS.			
	Discharged Recovered, Relieved, or Unimproved.*		Yorkshire (County and Borough Asylums).		England and Wales (County and Borough Asylums).	
Years.	To each 100 Admissions.	Increase between the two periods.	To each 100 Resident.	Decrease between the two periods.	To each 100 Resident.	Decrease between the two periods.
1818-67	50·25	} 7·05	13·87	} 2·45		
1868-77	57·30		11·42		10·60	} 0·61
1878-87	58·60	} 1·30	11·04	} 0·38	9·99	

* As the discharged "unimproved" include the transfers to other Asylums in the same County, the value of this table is very small so far as the Discharges are concerned, in its bearing on the accumulation of cases, and, therefore, the question of the increase of occurring insanity.

TABLE XVI.—Showing the Total Admissions, Discharges, and Deaths of Pauper Lunatics in Yorkshire from 1858 to 1887, in Decennial Periods (County and Borough Asylums only).

Periods.	Total Admissions.	Discharges and Deaths.			Excess of Admissions over Discharges and Deaths.	
		Discharges.	Deaths.	Discharges and Deaths.	Total Number.	Average Annual Number.
1858-67	5,474	2,838	2,165	5,003	471	47·1
1868-77	9,548	5,328	2,894	8,222	1,326	132·6
1878-87	12,331	7,201	4,097	11,298	1,033	103·3
1858-87	27,353	15,367	9,156	24,523	2,830	94·3

TABLE XVII.—Showing the total number of Pauper Lunatics chargeable to the Unions of Yorkshire* on Jan. 1st in the following years, and the ratio to the population in the County; also the increase in the population of the County and of Pauper Lunatics during the decades 1861-71, 1871-81, and 1881-89.

Years.	Population (enumerated at each Census).	Pauper Lunatics.	Numbers to Population.	Increase of *General* Population (Per Cent.).	Increase of *Lunatic* Population (Per Cent.).
1861	2,015,541	2647	1 in 761		40·4
1871	2,395,569	3717	1 in 644	18·85	33·36
1881	2,894,759	4957	1 in 584	20·83	
(8 years) 1889	3,310,252	5911	1 in 560	14·35	17·49
England & Wales 1889	29,015,613	75,632	1 in 384		

* In the York Union, embracing an area of 266 square miles, there were in 1840, 34 insane paupers in the city population, or 1 to 882. In the rural population of the same Union there were 23 insane paupers, or 1 to 565. In all there were 57, or 1 to 754.

TABLE XVIII.—Showing the ratio of the total number of Pauper Lunatics to 10,000 of the population of Yorkshire, 1861, 1871, 1881, and 1889, distinguishing those in Asylums, Workhouses, and Boarded-Out or with Relatives.

Years.	In Asylums.	In Workhouses.	With Friends, etc.	Total.
1861	7·80	3·45	1·88	13·13
1871	9·40	4·43	1·68	15·51
1881	11·93	4·40	1·04	17·37
1889	12·88	3·98	0·99	17·85

N.B.—In England and Wales the ratio of Pauper Lunatics to 10,000 of the population was, on the 1st of Jan., 1889, 26·06. In Durham the ratio is lower than in Yorkshire, namely, 14·96. In Middlesex it is as high as 38·35.

TABLE **XIX.**—Showing the number of Re-admissions and Transfers into the County and Borough Asylums of Yorkshire to January 1st, 1888, with their Percentages on the Total number of Cases Admitted during the Periods specified.

Asylums.	Re-admissions.	Total Admissions (Cases).	Persons Admitted.		Percentages.	
			Including Transfers.	Excluding Transfers.	Re-admissions to Total Admissions.	Transfers to Total Admissions.
West Riding Asylum (Wakefield), 1873 to 1887.*	1,000	7,248	6,248	6,165	13·80	1·14
West Riding Asylum (Wadsley), 1872 to 1887...	857	6,929	6,072	5,061	12·37	14·59
North Riding Asylum, near York, 1847 to 1887...........	643	5,155	4,512	4,098	12·47	8·03
East Riding Asylum (Beverly), 1871 to 1887.....	136	1,313	1,177	859	10·35	24·2
Hull Borough Asylum, 1849 to 1887............................	355	2,199	1,844	1,808	16·14	1·63
	2,991	22,844	19,853	17,991	13·09	8·15
Total Number of Transfers.			1,862.			

* Transfers and Re-admissions are not distinguished at an earlier date at the Wakefield Asylum than 1873.

SELECTION

FROM

J. & A. CHURCHILL'S GENERAL CATALOGUE

COMPRISING

ALL RECENT WORKS PUBLISHED BY THEM

ON THE

ART AND SCIENCE OF MEDICINE

N.B.—As far as possible, this List is arranged in the order in which medical study is usually pursued.

J. & A. CHURCHILL publish for the following Institutions and Public Bodies :—

ROYAL COLLEGE OF SURGEONS.
CATALOGUES OF THE MUSEUM.
Twenty-three separate Catalogues (List and Prices can be obtained of J. & A. CHURCHILL).

GUY'S HOSPITAL.
REPORTS BY THE MEDICAL AND SURGICAL STAFF.
Vol. XXX., Third Series. 7s. 6d.
FORMULÆ USED IN THE HOSPITAL IN ADDITION TO THOSE
IN THE B.P. 1s. 6d.

LONDON HOSPITAL.
PHARMACOPŒIA OF THE HOSPITAL. 3s.

ST. BARTHOLOMEW'S HOSPITAL.
CATALOGUE OF THE ANATOMICAL AND PATHOLOGICAL
MUSEUM. Vol. I.—Pathology. 15s. Vol. II.—Teratology, Anatomy
and Physiology, Botany. 7s. 6d.

ST. GEORGE'S HOSPITAL.
REPORTS BY THE MEDICAL AND SURGICAL STAFF.
The last Volume (X.) was issued in 1880. Price 7s. 6d.
CATALOGUE OF THE PATHOLOGICAL MUSEUM. 15s.
SUPPLEMENTARY CATALOGUE (1882). 5s.

ST. THOMAS'S HOSPITAL.
REPORTS BY THE MEDICAL AND SURGICAL STAFF
Annually. Vol. XVII., New Series. 7s. 6d.

MIDDLESEX HOSPITAL.
CATALOGUE OF THE PATHOLOGICAL MUSEUM. 12s.

WESTMINSTER HOSPITAL.
REPORTS BY THE MEDICAL AND SURGICAL STAFF.
Annually. Vol. IV. 6s.

ROYAL LONDON OPHTHALMIC HOSPITAL.
REPORTS BY THE MEDICAL AND SURGICAL STAFF.
Occasionally. Vol. XII., Part III. 5s.

OPHTHALMOLOGICAL SOCIETY OF THE UNITED KINGDOM
TRANSACTIONS. Vol. VIII. 12s. 6d.

MEDICO-PSYCHOLOGICAL ASSOCIATION.
JOURNAL OF MENTAL SCIENCE. Quarterly. 3s. 6d.

PHARMACEUTICAL SOCIETY OF GREAT BRITAIN.
PHARMACEUTICAL JOURNAL AND TRANSACTIONS.
Every Saturday. 4d. each, or 20s. per annum, post free.

BRITISH PHARMACEUTICAL CONFERENCE.
YEAR BOOK OF PHARMACY. 10s.

BRITISH DENTAL ASSOCIATION.
JOURNAL OF THE ASSOCIATION AND MONTHLY REVIEW
OF DENTAL SURGERY.
On the 15th of each Month. 6d. each, or 7s. per annum, post free.

A SELECTION

FROM

J. & A. CHURCHILL'S GENERAL CATALOGUE,

COMPRISING

ALL RECENT WORKS PUBLISHED BY THEM ON THE ART AND SCIENCE OF MEDICINE.

N.B.—*J. & A. Churchill's Descriptive List of Works on Chemistry, Materia Medica, Pharmacy, Botany, Photography, Zoology, the Microscope, and other Branches of Science, can be had on application.*

Practical Anatomy :
A Manual of Dissections. By CHRISTOPHER HEATH, Surgeon to University College Hospital. Seventh Edition. Revised by RICKMAN J. GODLEE, M.S. Lond., F.R.C.S., Teacher of Operative Surgery, late Demonstrator of Anatomy in University College, and Surgeon to the Hospital. Crown 8vo, with 24 Coloured Plates and 278 Engravings, 15s.

Wilson's Anatomist's Vade-Mecum. Tenth Edition. By GEORGE BUCHANAN, Professor of Clinical Surgery in the University of Glasgow ; and HENRY E. CLARK, M.R.C.S., Lecturer on Anatomy at the Glasgow Royal Infirmary School of Medicine. Crown 8vo, with 450 Engravings (including 26 Coloured Plates), 18s.

Braune's Atlas of Topographical Anatomy, after Plane Sections of Frozen Bodies. Translated by EDWARD BELLAMY, Surgeon to, and Lecturer on Anatomy, &c., at, Charing Cross Hospital. Large Imp. 8vo, with 34 Photolithographic Plates and 46 Woodcuts, 40s.

An Atlas of Human Anatomy. By RICKMAN J. GODLEE, M.S., F.R.C.S., Assistant Surgeon and Senior Demonstrator of Anatomy, University College Hospital. With 48 Imp. 4to Plates (112 figures), and a volume of Explanatory Text. 8vo, £4 14s. 6d.

Harvey's (Wm.) Manuscript Lectures. Prelectiones Anatomiæ Universalis. Edited, with an Autotype reproduction of the Original, by a Committee of the Royal College of Physicians of London. Crown 4to, half bound in Persian, 52s. 6d.

Anatomy of the Joints of Man. By HENRY MORRIS, Surgeon to, and Lecturer on Anatomy and Practical Surgery at, the Middlesex Hospital. 8vo, with 44 Lithographic Plates (several being coloured) and 13 Wood Engravings, 16s.

Manual of the Dissection of the Human Body. By LUTHER HOLDEN, Consulting Surgeon to St. Bartholomew's Hospital. Edited by JOHN LANGTON, F.R.C.S., Surgeon to, and Lecturer on Anatomy at, St. Bartholomew's Hospital. Fifth Edition. 8vo, with 208 Engravings. 20s.

By the same Author.

Human Osteology.
Seventh Edition, edited by CHARLES STEWART, Conservator of the Museum R.C.S., and R.W. REID, M.D., F.R.C.S., Lecturer on Anatomy at St. Thomas's Hospital. 8vo, with 59 Lithographic Plates and 75 Engravings. 16s.

Also.

Landmarks, Medical and Surgical. Fourth Edition. 8vo, 3s. 6d.

The Student's Guide to Surgical Anatomy. By EDWARD BELLAMY, F.R.C.S. and Member of the Board of Examiners. Third Edition. Fcap. 8vo, with 81 Engravings. 7s. 6d.

Diagrams of the Nerves of the Human Body, exhibiting their Origin, Divisions, and Connections, with their Distribution to the Various Regions of the Cutaneous Surface, and to all the Muscles. By W. H. FLOWER, C.B., F.R.S., F.R.C.S. Third Edition, with 6 Plates. Royal 4to, 12s.

General Pathology :

An Introduction to. By JOHN BLAND SUTTON, F.R.C.S., Sir E. Wilson Lecturer on Pathology, R.C.S. ; Assistant Surgeon to, and Lecturer on Anatomy at, Middlesex Hospital. 8vo, with 149 Engravings, 14s.

Atlas of Pathological Anatomy.

By Dr. LANCEREAUX. Translated by W. S. GREENFIELD, M.D., Professor of Pathology in the University of Edinburgh. Imp. 8vo, with 70 Coloured Plates, £5 5s.

A Manual of Pathological Anatomy.

By C. HANDFIELD JONES, M.B., F.R.S., and E. H. SIEVEKING, M.D., F.R.C.P. Edited by J. F. PAYNE, M.D., F.R.C.P., Lecturer on General Pathology at St. Thomas's Hospital. Second Edition. Crown 8vo, with 195 Engravings, 16s.

Post-mortem Examinations :

A Description and Explanation of the Method of Performing them, with especial reference to Medico-Legal Practice. By Prof. VIRCHOW. Translated by Dr. T. P. SMITH. Second Edition. Fcap. 8vo, with 4 Plates, 3s. 6d.

The Human Brain :

Histological and Coarse Methods of Research. A Manual for Students and Asylum Medical Officers. By W. BEVAN LEWIS, L.R.C.P. Lond., Medical Superintendent; West Riding Lunatic Asylum. 8vo, with Wood Engravings and Photographs, 8s.

Manual of Physiology :

For the use of Junior Students of Medicine. By GERALD F. YEO, M.D., F.R.C.S., F.R.S., Professor of Physiology in King's College, London. Second Edition. Crown 8vo, with 318 Engravings, 14s.

Principles of Human Physiology.

By W. B. CARPENTER, C.B., M.D., F.R.S. Ninth Edition. By HENRY POWER, M.B., F.R.C.S. 8vo, with 3 Steel Plates and 377 Wood Engravings, 31s. 6d.

Elementary Practical Biology :

Vegetable. By THOMAS W. SHORE, M.D., B.Sc. Lond., Lecturer on Comparative Anatomy at St. Bartholomew's Hospital. 8vo, 6s.

A Text-Book of Medical Physics,

for Students and Practitioners. By J. C. DRAPER, M.D., LL.D., Professor of Physics in the University of New York. With 377 Engravings. 8vo, 18s.

Medical Jurisprudence :

Its Principles and Practice. By ALFRED S. TAYLOR, M.D., F.R.C.P., F.R.S. Third Edition, by THOMAS STEVENSON, M.D., F.R.C.P., Lecturer on Medica Jurisprudence at Guy's Hospital. 2 vols. 8vo, with 188 Engravings, 31s. 6d.

By the same Authors.

A Manual of Medical Jurisprudence.

Eleventh Edition. Crown 8vo, with 56 Engravings, 14s.

Also.

Poisons,

In Relation to Medical Jurisprudence and Medicine. Third Edition. Crown 8vo, with 104 Engravings, 16s.

Lectures on Medical Jurisprudence.

By FRANCIS OGSTON, M.D., late Professor in the University of Aberdeen. Edited by FRANCIS OGSTON, Jun., M.D. 8vo, with 12 Copper Plates, 18s.

The Student's Guide to Medical Jurisprudence.

By JOHN ABERCROMBIE, M.D., F.R.C.P., Lecturer on Forensic Medicine to Charing Cross Hospital. Fcap. 8vo, 7s. 6d.

Microscopical Examination of Drinking Water and of Air.

By J. D. MACDONALD, M.D., F.R.S., Ex Professor of Naval Hygiene in the Army Medical School. Second Edition. 8vo, with 25 Plates, 7s. 6d.

Pay Hospitals and Paying Wards throughout the World.

By HENRY C. BURDETT. 8vo, 7s.

Hospitals, Infirmaries, and Dispensaries :

Their Construction, Interior Arrangement, and Management; with Descriptions of existing Institutions, and 74 Illustrations. By F. OPPERT, M.D., M.R.C.P.L. Second Edition. Royal 8vo, 12s.

Hospital Construction and Management.

By F. J. MOUAT, M.D., Local Government Board Inspector, and H. SAXON SNELL, Fell. Roy. Inst. Brit. Architects. Half calf, with large Map, 54 Lithographic Plates, and 27 Woodcuts, 35s.

Public Health Reports.

By Sir JOHN SIMON, C.B., F.R.S. Edited by EDWARD SEATON, M.D., F.R.C.P. 2 vols. 8vo, with Portrait, 36s.

Sanitary Examinations

Of Water, Air, and Food. A Vade-Mecum for the Medical Officer of Health. By CORNELIUS B. FOX, M.D., F.R.C.P. Second Edition. Crown 8vo, with 110 Engravings, 12s. 6d.

A Manual of Practical Hygiene.
By E. A. PARKES, M.D., F.R.S. Seventh Edition, by F. DE CHAUMONT, M.D., F.R.S., Professor of Military Hygiene in the Army Medical School. 8vo, with 9 Plates and 101 Engravings, 18s.

A Handbook of Hygiene and Sanitary Science.
By GEO. WILSON, M.A., M.D., F.R.S.E., Medical Officer of Health for Mid-Warwickshire. Sixth Edition. Crown 8vo, with Engravings. 10s. 6d.

By the same Author.

Healthy Life and Healthy Dwellings :
A Guide to Personal and Domestic Hygiene. Fcap. 8vo, 5s.

Epidemic Influences :
Epidemiological Aspects of Yellow Fever and of Cholera. The Milroy Lectures. By ROBERT LAWSON, LL.D., Inspector-General of Hospitals. 8vo, with Maps, Diagrams, &c., 6s.

Detection of Colour-Blindness and Imperfect Eyesight.
By CHARLES ROBERTS, F.R.C.S. Second Edition. 8vo, with a Table of Coloured Wools, and Sheet of Test-types, 5s.

Illustrations of the Influence of the Mind upon the Body in Health and Disease :
Designed to elucidate the Action of the Imagination. By D. H. TUKE, M.D., F.R.C.P., LL.D. Second Edition. 2 vols. crown 8vo, 15s.

By the same Author.

Sleep-Walking and Hypnotism.
8vo, 5s.

A Manual of Psychological Medicine.
With an Appendix of Cases. By JOHN C. BUCKNILL, M.D., F.R.S., and D. HACK TUKE, M.D., F.R.C.P. Fourth Edition. 8vo, with 12 Plates (30 Figures) and Engravings, 25s.

Mental Affections of Childhood and Youth
(Lettsomian Lectures for 1887, &c.). By J. LANGDON DOWN, M.D., F.R.C.P., Senior Physician to the London Hospital. 8vo, 6s.

Mental Diseases :
Clinical Lectures. By T. S. CLOUSTON, M.D., F.R.C.P. Edin., Lecturer on Mental Diseases in the University of Edinburgh. Second Edition. Crown 8vo, with 8 Plates (6 Coloured), 12s. 6d.

Intra-Uterine Death :
(Pathology of). Being the Lumleian Lectures, 1887. By WILLIAM O. PRIESTLEY, M.D., F.R.C.P., LL.D., Consulting Physician to King's College Hospital. 8vo, with 3 Coloured Plates and 17 Engravings, 7s. 6d.

Manual of Midwifery.
By ALFRED L. GALABIN, M.A., M.D., F.R.C.P., Obstetric Physician to, and Lecturer on Midwifery, &c. at, Guy's Hospital. Crown 8vo, with 227 Engravings, 15s.

The Student's Guide to the Practice of Midwifery.
By D. LLOYD ROBERTS, M.D., F.R.C.P., Lecturer on Clinical Midwifery and Diseases of Women at the Owens College ; Obstetric Physician to the Manchester Royal Infirmary. Third Edition. Fcap. 8vo, with 2 Coloured Plates and 127 Wood Engravings, 7s. 6d.

Lectures on Obstetric Operations :
Including the Treatment of Hæmorrhage, and forming a Guide to the Management of Difficult Labour. By ROBERT BARNES, M.D., F.R.C.P., Consulting Obstetric Physician to St. George's Hospital. Fourth Edition. 8vo, with 121 Engravings, 12s. 6d.

By the same Author.

A Clinical History of Medical and Surgical Diseases of Women.
Second Edition. 8vo, with 181 Engravings, 28s.

Clinical Lectures on Diseases of Women :
Delivered in St. Bartholomew's Hospital, by J. MATTHEWS DUNCAN, M.D., LL.D., F.R.S. Third Edition. 8vo, 16s.

The Female Pelvic Organs :
Their Surgery, Surgical Pathology, and Surgical Anatomy. In a Series of Coloured Plates taken from Nature ; with Commentaries, Notes, and Cases. By HENRY SAVAGE, M.D., F.R.C.S., Consulting Officer of the Samaritan Free Hospital. Fifth Edition. Roy. 4to, with 17 Lithographic Plates (15 coloured) and 52 Woodcuts, £1 15s.

Notes on Diseases of Women :
Specially designed to assist the Student in preparing for Examination. By J. J. REYNOLDS, L.R.C.P., M.R.C.S. Third Edition. Fcap. 8vo, 2s. 6d.

By the same Author.

Notes on Midwifery :
Specially designed for Students preparing for Examination. Second Edition. Fcap 8vo, with 15 Engravings, 4s.

A Manual of Obstetrics.
By A. F. A. KING, A.M., M.D., Professor of Obstetrics, &c., in the Columbian University, Washington, and the University of Vermont. Third Edition. Crown 8vo, with 102 Engravings, 8s.

The Student's Guide to the Diseases of Women.
By ALFRED L. GALABIN, M.D., F.R.C.P., Obstetric Physician to Guy's Hospital. Fourth Edition. Fcap. 8vo, with 94 Engravings, 7s. 6d.

West on the Diseases of Women.
Fourth Edition, revised by the Author, with numerous Additions by J. MATTHEWS DUNCAN, M.D., F.R.C.P., F.R.S.E., Obstetric Physician to St. Bartholomew's Hospital. 8vo, 16s.

Obstetric Aphorisms :
For the Use of Students commencing Midwifery Practice. By JOSEPH G. SWAYNE, M.D. Ninth Edition. Fcap. 8vo, with 17 Engravings, 3s. 6d.

Handbook of Midwifery for Midwives:
By J. E. BURTON, L.R.C.P. Lond., Surgeon to the Hospital for Women, Liverpool. Second Edition. With Engravings. Fcap. 8vo, 6s.

A Handbook of Uterine Therapeutics,
and of Diseases of Women. By E. J. TILT, M.D., M.R.C.P. Fourth Edition. Post 8vo, 10s.

By the same Author.

The Change of Life
In Health and Disease : A Clinical Treatise on the Diseases of the Nervous System incidental to Women at the Decline of Life. Fourth Edition. 8vo, 10s. 6d.

Diseases of the Uterus, Ovaries, and Fallopian Tubes :
A Practical Treatise by A. COURTY, Professor of Clinical Surgery, Montpellier. Translated from Third Edition by his Pupil, AGNES McLAREN, M.D., M.K.Q.C.P.I., with Preface by J. MATTHEWS DUNCAN, M.D., F.R.C.P. 8vo, with 424 Engravings, 24s.

Gynæcological Operations :
(Handbook of). By ALBAN H. G. DORAN, F.R.C.S., Surgeon to the Samaritan Hospital. 8vo, with 167 Engravings, 15s.

Diseases and Accidents
Incident to Women, and the Practice of Medicine and Surgery applied to them. By W. H. BYFORD, A.M., M.D., Professor of Gynæcology in Rush Medical College, and HENRY T. BYFORD, M.D., Surgeon to the Woman's Hospital, Chicago. Fourth Edition. 8vo, with 306 Engravings, 25s.

A Practical Treatise on the Diseases of Women.
By T. GAILLARD THOMAS, M.D., Professor of Diseases of Women in the College of Physicians and Surgeons, New York. Fifth Edition. Roy. 8vo, with 266 Engravings, 25s.

Abdominal Surgery.
By J. GREIG SMITH, M.A., F.R.S.E., Surgeon to the Bristol Royal Infirmary and Lecturer on Surgery in the Bristol Medical School. Second Edition. 8vo, with 79 Engravings, 21s.

The Student's Guide to Diseases of Children.
By JAS. F. GOODHART, M.D., F.R.C.P., Physician to Guy's Hospital, and to the Evelina Hospital for Sick Children. Third Edition. Fcap. 8vo, 10s. 6d.

Diseases of Children.
For Practitioners and Students. By W. H. DAY, M.D., Physician to the Samaritan Hospital. Second Edition. Crown 8vo, 12s. 6d.

A Practical Treatise on Disease in Children.
By EUSTACE SMITH, M.D., Physician to the King of the Belgians, Physician to the East London Hospital for Children. 8vo, 22s.

By the same Author.

Clinical Studies of Disease in Children.
Second Edition. Post 8vo, 7s. 6d. *Also.*

The Wasting Diseases of Infants and Children.
Fifth Edition. Post 8vo, 8s. 6d.

A Practical Manual of the Diseases of Children.
With a Formulary. By EDWARD ELLIS, M.D. Fifth Edition. Crown 8vo, 10s.

A Manual for Hospital Nurses
and others engaged in Attending on the Sick, and a Glossary. By EDWARD J. DOMVILLE, Surgeon to the Exeter Lying-in Charity. Sixth Edition. Cr. 8vo, 2s. 6d.

A Manual of Nursing, Medical and Surgical.
By CHARLES J. CULLINGWORTH, M.D., Obstetric Physician to St. Thomas's Hospital. Second Edition. Fcap. 8vo, with Engravings, 3s. 6d.

By the same Author.

A Short Manual for Monthly Nurses.
Second Edition. Fcap. 8vo, 1s. 6d.

Hospital Sisters and their Duties.
By EVA C. E. LÜCKES, Matron to the London Hospital. Second Edition. Crown 8vo, 2s. 6d.

Diseases and their Commencement.
Lectures to Trained Nurses. By DONALD W. C. HOOD, M.D., M.R.C.P., Physician to the West London Hospital. Crown 8vo, 2s. 6d.

Infant Feeding and its Influence on Life.
By C. H. F. ROUTH, M.D., Physician to the Samaritan Hospital. Fourth Edition. Fcap. 8vo. [*Preparing.*

Manual of Botany:

Including the Structure, Classification, Properties, Uses, and Functions of Plants. By ROBERT BENTLEY, Professor of Botany in King's College and to the Pharmaceutical Society. Fifth Edition. Crown 8vo, with 1,178 Engravings, 15s.

By the same Author.

The Student's Guide to Structural, Morphological, and Physiological Botany.

With 660 Engravings. Fcap. 8vo, 7s. 6d.

Also.

The Student's Guide to Systematic Botany,

including the Classification of Plants and Descriptive Botany. Fcap. 8vo, with 350 Engravings, 3s. 6d.

Medicinal Plants:

Being descriptions, with original figures, of the Principal Plants employed in Medicine, and an account of their Properties and Uses. By Prof. BENTLEY and Dr. H. TRIMEN, F.R.S. In 4 vols., large 8vo, with 306 Coloured Plates, bound in Half Morocco, Gilt Edges, £11 11s.

Outlines of Infectious Diseases:

For the use of Clinical Students. By J. W. ALLAN, M.B., Physician Superintendent Glasgow Fever Hospital. Fcap. 8vo., 3s.

Materia Medica.

A Manual for the use of Students. By ISAMBARD OWEN, M.D., F.R.C.P., Lecturer on Materia Medica, &c., to St. George's Hospital. Second Edition. Crown 8vo, 6s. 6d.

The Prescriber's Pharmacopœia:

The Medicines arranged in Classes according to their Action, with their Composition and Doses. By NESTOR J. C. TIRARD, M.D., F.R.C.P., Professor of Materia Medica and Therapeutics in King's College, London. Sixth Edition. 32mo, bound in leather, 3s.

Royle's Manual of Materia Medica and Therapeutics.

Sixth Edition, including additions and alterations in the B.P. 1885. By JOHN HARLEY, M.D., Physician to St. Thomas's Hospital. Crown 8vo, with 139 Engravings, 15s.

Materia Medica and Therapeutics:

Vegetable Kingdom — Organic Compounds — Animal Kingdom. By CHARLES D. F. PHILLIPS, M.D., F.R.S. Edin., late Lecturer on Materia Medica and Therapeutics at the Westminster Hospital Medical School. 8vo, 25s.

The Student's Guide to Materia Medica and Therapeutics.

By JOHN C. THOROWGOOD, M.D., F.R.C.P. Second Edition. Fcap. 8vo, 7s.

A Companion to the British Pharmacopœia.

By PETER SQUIRE, Revised by his Sons, P. W. and A. H. SQUIRE. 14th Edition. 8vo, 10s. 6d.

By the same Authors.

The Pharmacopœias of the London Hospitals,

arranged in Groups for Easy Reference and Comparison. Fifth Edition 18mo, 6s.

A Treatise on the Principles and Practice of Medicine.

Sixth Edition. By AUSTIN FLINT, M.D., W.H. WELCH, M.D., and AUSTIN FLINT, jun., M.D. 8vo, with Engravings, 26s.

Climate and Fevers of India,

with a series of Cases (Croonian Lectures, 1882). By Sir JOSEPH FAYRER, K.C.S.I., M.D. 8vo, with 17 Temperature Charts, 12s.

By the same Author.

The Natural History and Epidemiology of Cholera:

Being the Annual Oration of the Medical Society of London, 1888. 8vo, 3s. 6d.

Family Medicine and Hygiene for India.

A Manual. By Sir WILLIAM J. MOORE, M.D., K.C.I.E., late Surgeon-General with the Government of Bombay. Published under the Authority of the Government of India. Fifth Edition. Post 8vo, with 71 Engravings, 12s.

By the same Author.

A Manual of the Diseases of India:

With a Compendium of Diseases generally. Second Edition. Post 8vo, 10s.

The Prevention of Disease in Tropical and Sub-Tropical Campaigns.

(Parkes Memorial Prize for 1886.) By ANDREW DUNCAN, M.D., B.S. Lond., F.R.C.S., Surgeon, Bengal Army. 8vo, 12s. 6d.

Practical Therapeutics:

A Manual. By EDWARD J. WARING, C.I.E., M.D., F.R.C.P., and DUDLEY W. BUXTON, M.D., B.S. Lond. Fourth Edition. Crown 8vo, 14s.

By the same Author.

Bazaar Medicines of India,

And Common Medical Plants: With Full Index of Diseases, indicating their Treatment by these and other Agents procurable throughout India, &c. Fourth Edition Fcap 8vo, 5s.

A Commentary on the Diseases of India.

By NORMAN CHEVERS, C.I.E., M.D., F.R.C.S., Deputy Surgeon-General H.M. Indian Army. 8vo, 24s.

The Principles and Practice of Medicine.
By C. HILTON FAGGE, M.D. Second Edition. Edited by P. H. PYE-SMITH, M.D., F.R.S., F.R.C.P., Physician to, and Lecturer on Medicine in, Guy's Hospital. 2 vols. 8vo. Cloth, 38s. ; Half Leather, 44s.

The Student's Guide to the Practice of Medicine.
By M. CHARTERIS, M.D., Professor of Therapeutics and Materia Medica in the University of Glasgow. Fifth Edition. Fcap. 8vo, with Engravings on Copper and Wood, 9s.

Hooper's Physicians' Vade-Mecum.
A Manual of the Principles and Practice of Physic. Tenth Edition. By W. A. GUY, F.R.C.P., F.R.S., and J. HARLEY, M.D., F.R.C.P. With 118 Engravings. Fcap. 8vo, 12s. 6d.

Preventive Medicine.
Collected Essays. By WILLIAM SQUIRE, M.D., F.R.C.P., Physician to St. George, Hanover-square, Dispensary. 8vo, 6s. 6d.

The Student's Guide to Clinical Medicine and Case-Taking.
By FRANCIS WARNER, M.D., F.R.C.P., Physician to the London Hospital. Second Edition. Fcap. 8vo, 5s.

An Atlas of the Pathological Anatomy of the Lungs.
By the late WILSON FOX, F.R.C.P., Physician to H.M. the Queen. With 45 Plates (mostly Coloured) and Engravings. 4to, half-bound in Calf, 70s.

The Bronchi and Pulmonary Blood-vessels :
their Anatomy and Nomenclature. By WILLIAM EWART, M.D., F.R.C.P., Physician to St. George's Hospital. 4to, with 20 Illustrations, 21s.

The Student's Guide to Diseases of the Chest.
By VINCENT D. HARRIS, M.D. Lond., F.R.C.P., Physician to the City of London Hospital for Diseases of the Chest, Victoria Park. Fcap. 8vo, with 55 Illustrations (some Coloured), 7s. 6d.

How to Examine the Chest :
A Practical Guide for the use of Students. By SAMUEL WEST, M.D., F.R.C.P., Physician to the City of London Hospital for Diseases of the Chest; Assistant Physician to St. Bartholomew's Hospital. With 42 Engravings. Fcap. 8vo, 5s.

Contributions to Clinical and Practical Medicine.
By A. T. HOUGHTON WATERS, M.D., Physician to the Liverpool Royal Infirmary. 8vo, with Engravings, 7s.

Fever : A Clinical Study.
By T. J. MACLAGAN, M.D. 8vo, 7s. 6d.

The Student's Guide to Medical Diagnosis.
By SAMUEL FENWICK, M.D., F.R.C.P., Physician to the London Hospital, and BEDFORD FENWICK, M.D., M.R.C.P. Sixth Edition. Fcap. 8vo, with 114 Engravings, 7s.

By the same Author.

The Student's Outlines of Medical Treatment.
Second Edition. Fcap. 8vo, 7s.

Also.

On Chronic Atrophy of the Stomach,
and on the Nervous Affections of the Digestive Organs. 8vo, 8s.

Also.

The Saliva as a Test for Functional Diseases of the Liver.
Crown 8vo, 2s.

The Microscope in Medicine.
By LIONEL S. BEALE, M.B., F.R.S., Physician to King's College Hospital. Fourth Edition. 8vo, with 86 Plates, 21s.

Also.

On Slight Ailments :
Their Nature and Treatment. Second Edition. 8vo, 5s.

Medical Lectures and Essays.
By G. JOHNSON, M.D., F.R.C.P., F.R.S., Consulting Physician to King's College Hospital. 8vo, with 46 Engravings, 25s.

By the same Author.

An Essay on Asphyxia (Apnœa).
8vo, 3s.

Winter Cough
(Catarrh, Bronchitis, Emphysema, Asthma). By HORACE DOBELL, M.D., Consulting Physician to the Royal Hospital for Diseases of the Chest. Third Edition. 8vo, with Coloured Plates, 10s. 6d.

By the same Author.

Loss of Weight, Blood-Spitting, and Lung Disease.
Second Edition. 8vo, with Chromo-lithograph, 10s. 6d.

Also.

The Mont Dore Cure, and the Proper Way to Use it.
8vo, 7s. 6d.

Vaccinia and Variola :
A Study of their Life History. By JOHN B. BUIST, M.D., F.R.S.E., Teacher of Vaccination for the Local Government Board. Crown 8vo, with 24 Coloured Plates, 7s. 6d.

Treatment of Some of the Forms of Valvular Disease of the Heart.
By A. E. SANSOM, M.D., F.R.C.P., Physician to the London Hospital. Second Edition. Fcap. 8vo, with 26 Engravings, 4s. 6d.

Diseases of the Heart and Aorta :

Clinical Lectures. By G. W. BALFOUR, M.D., F.R.C.P., F.R.S. Edin., late Senior Physician and Lecturer on Clinical Medicine, Royal Infirmary, Edinburgh. Second Edition. 8vo, with Chromo-lithograph and Wood Engravings, 12s. 6d.

Notes on Asthma :

Its Forms and Treatment. By JOHN C. THOROWGOOD, M.D., Physician to the Hospital for Diseases of the Chest Third Edition. Crown 8vo, 4s. 6d.

Medical Ophthalmoscopy :

A Manual and Atlas. By W. R. GOWERS, M.D., F.R.C.P., F.R.S., Physician to the National Hospital for the Paralyzed and Epileptic. Second Edition, with Coloured Plates and Woodcuts. 8vo, 18s.

By the same Author.

Diagnosis of Diseases of the Brain.
Second Edition. 8vo, with Engravings, 7s. 6d.

Also.

Diagnosis of Diseases of the Spinal Cord.
Third Edition. 8vo, with Engravings, 4s. 6d.

Also.

A Manual of Diseases of the Nervous System.

Vol. I. Diseases of the Spinal Cord and Nerves. Roy. 8vo, with 171 Engravings (many figures), 12s. 6d.

Vol. II. Diseases of the Brain and Cranial Nerves : General and Functional Diseases of the Nervous System. 8vo, with 170 Engravings, 17s. 6d.

Diseases of the Nervous System.

Lectures delivered at Guy's Hospital. By SAMUEL WILKS, M.D., F.R.S. Second Edition. 8vo, 18s.

Gout in its Clinical Aspects.

By J. MORTIMER GRANVILLE, M.D. Crown 8vo, 6s.

Regimen to be adopted in Cases of Gout.
By WILHELM EBSTEIN, M.D., Professor of Clinical Medicine in Göttingen. Translated by JOHN SCOTT, M.A., M.B. 8vo, 2s. 6d.

Diseases of the Nervous System.

Clinical Lectures. By THOMAS BUZZARD, M.D., F.R.C.P., Physician to the National Hospital for the Paralysed and Epileptic. With Engravings, 8vo. 15s.

By the same Author.

Some Forms of Paralysis from Peripheral Neuritis :
of Gouty, Alcoholic, Diphtheritic, and other origin. Crown 8vo, 5s.

Diseases of the Liver :

With and without Jaundice. By GEORGE HARLEY, M.D., F.R.C.P., F.R.S. 8vo, with 2 Plates and 36 Engravings, 21s.

By the same Author.

Inflammations of the Liver, and their Sequelæ.
Crown 8vo, with Engravings, 5s.

Gout, Rheumatism,

And the Allied Affections ; with Chapters on Longevity and Sleep. By PETER HOOD, M.D. Third Edition. Crown 8vo, 7s. 6d.

Diseases of the Stomach :

The Varieties of Dyspepsia, their Diagnosis and Treatment. By S. O. HABERSHON, M.D., F.R.C.P. Third Edition. Crown 8vo, 5s.

By the same Author.

Pathology of the Pneumogastric Nerve :
Lumleian Lectures for 1876. Second Edition. Post 8vo, 4s.

Also.

Diseases of the Abdomen,

Comprising those of the Stomach and other parts of the Alimentary Canal, Œsophagus, Cæcum, Intestines, and Peritoneum. Fourth Edition. 8vo, with 5 Plates, 21s.

Also.

Diseases of the Liver,

Their Pathology and Treatment. Lettsomian Lectures. Second Edition. Post 8vo, 4s.

On the Relief of Excessive and Dangerous Tympanites by Puncture of the Abdomen.
By JOHN W. OGLE, M.A., M.D., F.R.C.P., Consulting Physician to St. George's Hospital. 8vo, 5s. 6d.

Acute Intestinal Strangulation,

And Chronic Intestinal Obstruction (Mode of Death from). By THOMAS BRYANT, F.R.C.S., Senior Surgeon to Guy's Hospital. 8vo, 3s.

Handbook of the Diseases of the Nervous System.
By JAMES ROSS, M.D., F.R.C.P., Assistant Physician to the Manchester Royal Infirmary. Roy. 8vo, with 184 Engravings, 18s.

Also.

Aphasia :

Being a Contribution to the Subject of the Dissolution of Speech from Cerebral Disease. 8vo, with Engravings, 4s. 6d.

Food and Dietetics,

Physiologically and Therapeutically Considered. By F. W. PAVY, M.D., F.R.S., Physician to Guy's Hospital. Second Edition. 8vo, 15s.

By the same Author.

Croonian Lectures on Certain

Points connected with Diabetes. 8vo, 4s. 6d.

Headaches :

Their Nature, Causes, and Treatment. By W. H. DAY, M.D., Physician to the Samaritan Hospital. Fourth Edition. Crown 8vo, with Engravings, 7s. 6d.

Health Resorts at Home and

Abroad. By M. CHARTERIS, M.D., Professor of Therapeutics and Materia Medica in the University of Glasgow. Second Edition. Crown 8vo, with Map, 5s. 6d.

Winter and Spring

On the Shores of the Mediterranean. By HENRY BENNET, M.D. Fifth Edition. Post 8vo, with numerous Plates, Maps, and Engravings, 12s. 6d.

Medical Guide to the Mineral

Waters of France and its Wintering Stations. With a Special Map. By A. VINTRAS, M.D., Physician to the French Embassy, and to the French Hospital, London. Crown 8vo, 8s.

The Ocean as a Health-Resort :

A Practical Handbook of the Sea, for the use of Tourists and Health-Seekers. By WILLIAM S. WILSON, L.R.C.P. Second Edition, with Chart of Ocean Routes, &c. Crown 8vo, 7s. 6d.

Ambulance Handbook for Volun-

teers and Others. By J. ARDAVON RAYE, L.K. & Q.C.P.I., L.R.C.S.I., late Surgeon to H.B.M. Transport No. 14, Zulu Campaign, and Surgeon E.I.R. Rifles. 8vo, with 16 Plates (50 figures), 3s. 6d.

Ambulance Lectures :

To which is added a NURSING LECTURE. By JOHN M. H. MARTIN, Honorary Surgeon to the Blackburn Infirmary. Second Edition. Crown 8vo, with 59 Engravings, 2s.

Commoner Diseases and Acci-

dents to Life and Limb: their Prevention and Immediate Treatment. By M. M. BASIL, M.A., M.B., C.M. Crown 8vo, 2s. 6d.

How to Use a Galvanic Battery

in Medicine and Surgery. By HERBERT TIBBITS, M.D., F.R.C.P.E., Senior Physician to the West London Hospital for Paralysis and Epilepsy. Third Edition. 8vo, with Engravings, 4s.

By the same Author.

A Map of Ziemssen's Motor

Points of the Human Body : A Guide to Localised Electrisation. Mounted on Rollers, 35 × 21. With 20 Illustrations, 5s.

Also.

Electrical and Anatomical De-

monstrations. A Handbook for Trained Nurses and Masseuses. Crown 8vo, with 44 Illustrations, 5s.

Surgical Emergencies :

Together with the Emergencies attendant on Parturition and the Treatment of Poisoning. By W. PAUL SWAIN, F.R.C.S., Surgeon to the South Devon and East Cornwall Hospital. Fourth Edition. Crown 8vo, with 120 Engravings, 5s.

Operative Surgery in the Cal-

cutta Medical College Hospital. Statistics, Cases, and Comments. By KENNETH McLEOD, A.M., M.D., F.R.C.S.E., Surgeon-Major, Indian Medical Service, Professor of Surgery in Calcutta Medical College. 8vo, with Illustrations, 12s. 6d.

A Course of Operative Surgery.

By CHRISTOPHER HEATH, Surgeon to University College Hospital. Second Edition. With 20 coloured Plates (180 figures) from Nature, by M. LÉVEILLÉ, and several Woodcuts. Large 8vo, 30s.

By the same Author.

The Student's Guide to Surgical

Diagnosis. Second Edition. Fcap. 8vo, 6s. 6d.

Also.

Manual of Minor Surgery and

Bandaging. For the use of House-Surgeons, Dressers, and Junior Practitioners. Eighth Edition. Fcap. 8vo, with 142 Engravings, 6s.

Also.

Injuries and Diseases of the

Jaws. Third Edition. 8vo, with Plate and 206 Wood Engravings, 14s.

Also,

Lectures on Certain Diseases

of the Jaws. Delivered at the R.C.S., Eng., 1887. 8vo, with 64 Engravings, 2s. 6d.

The Practice of Surgery :

A Manual. By THOMAS BRYANT, Consulting Surgeon to Guy's Hospital. Fourth Edition. 2 vols. crown 8vo, with 750 Engravings (many being coloured), and including 6 chromo plates, 32s.

By the same Author.

On Tension : Inflammation of Bone, and Head Injuries. Hunterian Lectures, 1888. 8vo, 6s.

Surgery : its Theory and Practice (Student's Guide). By WILLIAM J. WALSHAM, F.R.C.S., Assistant Surgeon to St. Bartholomew's Hospital. Second Edition. Fcap. 8vo, with 294 Engravings, 10s. 6d.

The Surgeon's Vade-Mecum :

A Manual of Modern Surgery. By R. DRUITT, F.R.C.S. Twelfth Edition. By STANLEY BOYD, M.B., F.R.C.S. Assistant Surgeon and Pathologist to Charing Cross Hospital. Crown 8vo, with 373 Engravings 16s.

The Operations of Surgery :

Intended for Use on the Dead and Living Subject alike. By W. H. A. JACOBSON, M.A., M.B., M.Ch. Oxon., F.R.C.S., Assistant Surgeon to Guy's Hospital, and Teacher of Surgery in the Medical School. 8vo, with 200 Illustrations, 30s.

Regional Surgery :

Including Surgical Diagnosis. A Manual for the use of Students. By F. A. SOUTHAM, M.A., M.B., F.R.C.S., Assistant Surgeon to the Manchester Royal Infirmary. Part I. The Head and Neck. Crown 8vo, 6s. — Part II. The Upper Extremity and Thorax. Crown 8vo, 7s. 6d. Part III. The Abdomen and Lower Extremity. Crown 8vo, 7s.

A Treatise on Dislocations.

By LEWIS A. STIMSON, M.D., Professor of Clinical Surgery in the University of the City of New York. Roy. 8vo, with 163 Engravings, 15s.

By the same Author.

A Treatise on Fractures.

Roy. 8vo, with 360 Engravings, 21s.

Lectures on Orthopædic Surgery. By BERNARD E. BRODHURST, F.R.C.S., Surgeon to the Royal Orthopædic Hospital. Second Edition. 8vo, with Engravings, 12s. 6d.

By the same Author.

On Anchylosis, and the Treatment for the Removal of Deformity and the Restoration of Mobility in Various Joints. Fourth Edition. 8vo, with Engravings, 5s.

Also.

Curvatures and Disease of the Spine. Fourth Edition. 8vo, with Engravings, 7s. 6d.

Surgical Pathology and Morbid Anatomy (Student's Guide). By ANTHONY A. BOWLBY, F.R.C.S., Surgical Registrar and Demonstrator of Surgical Pathology to St. Bartholomew's Hospital. Fcap. 8vo, with 135 Engravings, 9s.

Illustrations of Clinical Surgery

By JONATHAN HUTCHINSON, F.R.S. Senior Surgeon to the London Hospital In fasciculi. 6s. 6d each. Fasc. I. to X. bound, with Appendix and Index, £3 10s. Fasc. XI. to XXIII. bound, with Index, £4 10s.

Diseases of Bones and Joints.

By CHARLES MACNAMARA, F.R.C.S., Surgeon to, and Lecturer on Surgery at, the Westminster Hospital. 8vo, with Plates and Engravings, 12s.

Injuries of the Spine and Spinal Cord, and NERVOUS SHOCK, in their Surgical and Medico-Legal Aspects. By HERBERT W. PAGE, M.C. Cantab., F.R.C.S., Surgeon to St. Mary's Hospital. Second Edition, post 8vo, 10s.

Spina Bifida :

Its Treatment by a New Method. By JAS. MORTON, M.D., L.R.C.S.E., Professor of Materia Medica in Anderson's College, Glasgow. 8vo, with Plates, 7s. 6d.

Face and Foot Deformities.

By FREDERICK CHURCHILL, C.M., Surgeon to the Victoria Hospital for Children. 8vo, with Plates and Illustrations, 10s. 6d.

Clubfoot :

Its Causes, Pathology, and Treatment. By WM. ADAMS, F.R.C.S., Surgeon to the Great Northern Hospital. Second Edition. 8vo, with 106 Engravings and 6 Lithographic Plates, 15s.

By the same Author.

Lateral and other Forms of Curvature of the Spine : Their Pathology and Treatment. Second Edition. 8vo, with 5 Lithographic Plates and 72 Wood Engravings, 10s. 6d.

Electricity and its Manner of Working in the Treatment of Disease. By WM. E. STEAVENSON, M.D., Physician and Electrician to St. Bartholomew's Hospital. 8vo, 4s. 6d.

On Diseases and Injuries of the Eye : A Course of Systematic and Clinical Lectures to Students and Medical Practitioners. By J. R. WOLFE, M.D., F.R.C.S.E., Lecturer on Ophthalmic Medicine and Surgery in Anderson's College, Glasgow. With 10 Coloured Plates and 157 Wood Engravings. 8vo, £1 1s.

Hints on Ophthalmic Out-Patient Practice. By CHARLES HIGGENS, Ophthalmic Surgeon to Guy's Hospital. Third Edition. Fcap. 8vo, 3s.

The Student's Guide to Diseases of the Eye.
By EDWARD NETTLESHIP, F.R.C.S., Ophthalmic Surgeon to St. Thomas's Hospital. Fourth Edition. Fcap. 8vo, with 164 Engravings and a Set of Coloured Papers illustrating Colour-Blindness, 7s. 6d.

Manual of the Diseases of the Eye.
By CHARLES MACNAMARA, F.R.C.S., Surgeon to Westminster Hospital. Fourth Edition. Crown 8vo, with 4 Coloured Plates and 66 Engravings, 10s. 6d.

Normal and Pathological Histology of the Human Eye and Eyelids.
By C. FRED. POLLOCK, M.D., F.R.C.S. and F,R.S.E., Surgeon for Diseases of the Eye to Anderson's College Dispensary, Glasgow. Crown 8vo, with 100 Plates (230 drawings), 15s.

Atlas of Ophthalmoscopy.
Composed of 12 Chromo-lithographic Plates (59 Figures drawn from nature) and Explanatory Text. By RICHARD LIEBREICH, M.R.C.S. Translated by H. ROSBOROUGH SWANZY, M.B. Third edition, 4to, 40s.

Refraction of the Eye:
A Manual for Students. By GUSTAVUS HARTRIDGE, F.R.C.S., Surgeon to the Royal Westminster Ophthalmic Hospital. Third Edition. Crown 8vo, with 96 Illustrations, Test-types, &c., 5s. 6d.

Squint:
(Clinical Investigations on). By C. SCHWEIGGER, M.D., Professor of Ophthalmology in the University of Berlin. Edited by GUSTAVUS HARTRIDGE, F.R.C.S., 8vo, 5s.

Practitioner's Handbook of Diseases of the Ear and Naso-Pharynx.
By H. MACNAUGHTON JONES, M.D., late Professor of the Queen's University in Ireland, Surgeon to the Cork Ophthalmic and Aural Hospital. Third Edition of "Aural Surgery." Roy. 8vo, with 128 Engravings, 6s.

By the same Author.

Atlas of Diseases of the Membrana Tympani.
In Coloured Plates, containing 62 Figures, with Text. Crown 4to, 21s.

Endemic Goitre or Thyreocele:
Its Etiology, Clinical Characters, Pathology, Distribution, Relations to Cretinism, Myxoedema, &c., and Treatment. By WILLIAM ROBINSON, M.D. 8vo, 5s.

Diseases and Injuries of the Ear.
By Sir WILLIAM B. DALBY, Aural Surgeon to St. George's Hospital. Third Edition. Crown 8vo, with Engravings, 7s. 6d.

By the Same Author.

Short Contributions to Aural Surgery, between 1875 and 1886.
8vo, with Engravings, 3s. 6d.

Diphtheria:
Its Nature and Treatment, Varieties, and Local Expressions. By Sir MORELL MACKENZIE, M.D., Senior Physician to the Hospital for Diseases of the Throat. 8vo, 5s.

Sore Throat:
Its Nature, Varieties, and Treatment. By PROSSER JAMES, M.D., Physician to the Hospital for Diseases of the Throat. Fifth Edition. Post 8vo, with Coloured Plates and Engravings, 6s. 6d.

Studies in Pathological Anatomy,
Especially in Relation to Laryngeal Neoplasms. By R. NORRIS WOLFENDEN, M.D., Senior Physician to the Throat Hospital, and SIDNEY MARTIN, M.D., Pathologist to the City of London Hospital, Victoria Park. I. Papilloma. Roy. 8vo, with Coloured Plates, 2s. 6d.

A System of Dental Surgery.
By Sir JOHN TOMES, F.R.S., and C. S. TOMES, M.A., F.R.S. Third Edition. Crown 8vo, with 292 Engravings, 15s.

Dental Anatomy, Human and Comparative:
A Manual. By CHARLES S. TOMES, M.A., F.R.S. Second Edition Crown 8vo, with 191 Engravings, 12s. 6d.

The Student's Guide to Dental Anatomy and Surgery.
By HENRY SEWILL, M.R.C.S., L.D.S. Second Edition. Fcap. 8vo, with 78 Engravings, 5s. 6d.

A Manual of Nitrous Oxide Anæsthesia, for the use of Students and General Practitioners.
By J. FREDERICK W. SILK, M.D. Lond., M.R.C.S., Anæsthetist to the Great Northern Central Hospital, and to the National Dental Hospital. 8vo, with 26 Engravings, 5s.

Mechanical Dentistry in Gold and Vulcanite.
By F. H. BALKWILL, L.D.S.R.C.S. 8vo, with 2 Lithographic Plates and 57 Engravings, 10s.

Principles and Practice of Dentistry : including Anatomy, Physiology, Pathology, Therapeutics, Dental Surgery, and Mechanism. By C. A. HARRIS, M.D., D.D.S. Edited by F. J. S. GORGAS, A.M., M.D., D.D.S., Professor in the Dental Department of Maryland University. Twelfth Edition. 8vo, with over 1,000 Illustrations, 33s.

A Practical Treatise on Mechanical Dentistry. By JOSEPH RICHARDSON, M.D., D.D.S., late Emeritus Professor of Prosthetic Dentistry in the Indiana Medical College. Fourth Edition. Roy. 8vo, with 458 Engravings, 21s.

Elements of Dental Materia Medica and Therapeutics, with Pharmacopœia. By JAMES STOCKEN, L.D.S.R.C.S., Pereira Prizeman for Materia Medica, and THOMAS GADDES, L.D.S. Eng. and Edin. Third Edition. Fcap. 8vo, 7s. 6d.

Papers on Dermatology. By E. D. MAPOTHER, M.D., Ex-Pres. R.C.S.I. 8vo, 3s. 6d.

Atlas of Skin Diseases. By TILBURY FOX, M.D., F.R.C.P. With 72 Coloured Plates. Royal 4to, half morocco, £6 6s.

Diseases of the Skin : With an Analysis of 8,000 Consecutive Cases and a Formulary. By L. D. BULKLEY, M.D., Physician for Skin Diseases at the New York Hospital. Crown 8vo, 6s. 6d.

By the same Author.

Acne : its Etiology, Pathology, and Treatment : Based upon a Study of 1,500 Cases. 8vo, with Engravings, 10s.

On Certain Rare Diseases of the Skin. By JONATHAN HUTCHINSON, F.R.S., Senior Surgeon to the London Hospital, and to the Hospital for Diseases of the Skin. 8vo, 10s. 6d.

Diseases of the Skin : A Practical Treatise for the Use of Students and Practitioners. By J. N. HYDE, A.M., M.D., Professor of Skin and Venereal Diseases, Rush Medical College, Chicago. Second Edition. 8vo, with 2 Coloured Plates and 96 Engravings, 20s.

Leprosy in British Guiana. By JOHN D. HILLIS, F.R.C.S., M.R.I.A., Medical Superintendent of the Leper Asylum, British Guiana. Imp. 8vo, with 22 Lithographic Coloured Plates and Wood Engravings, £1 11s. 6d.

On Cancer : Its Allies, and other Tumours ; their Medical and Surgical Treatment. By F. A. PURCELL, M.D., M.C., Surgeon to the Cancer Hospital, Brompton. 8vo, with 21 Engravings, 10s. 6d.

Sarcoma and Carcinoma : Their Pathology, Diagnosis, and Treatment. By HENRY T. BUTLIN, F.R.C.S., Assistant Surgeon to St. Bartholomew's Hospital. 8vo, with 4 Plates, 8s.

By the same Author.

Malignant Disease of the Larynx (Sarcoma and Carcinoma). 8vo, with 5 Engravings, 5s.

Also.

Operative Surgery of Malignant Disease. 8vo, 14s.

Cancerous Affections of the Skin. (Epithelioma and Rodent Ulcer.) By GEORGE THIN, M.D. Post 8vo, with 8 Engravings, 5s.

By the same Author.

Pathology and Treatment of Ringworm. 8vo, with 21 Engravings, 5s.

Cancer of the Mouth, Tongue, and Alimentary Tract : their Pathology, Symptoms, Diagnosis, and Treatment. By FREDERIC E. JESSETT, F.R.C.S., Surgeon to the Cancer Hospital, Brompton. 8vo, 10s.

Clinical Chemistry of Urine (Outlines of the). By C. A. MACMUNN, M.A., M.D. 8vo, with 70 Engravings. [*Nearly Ready.*]

Lectures on the Surgical Disorders of the Urinary Organs. By REGINALD HARRISON, F.R.C.S., Surgeon to the Liverpool Royal Infirmary. Third Edition, with 117 Engravings. 8vo, 12s. 6d.

Hydrocele : Its several Varieties and their Treatment. By SAMUEL OSBORN, late Surgical Registrar to St. Thomas's Hospital. Fcap. 8vo, with Engravings, 3s.

By the same Author.

Diseases of the Testis. Fcap. 8vo, with Engravings, 3s. 6d.

Diseases of the Urinary Organs.

Clinical Lectures. By Sir HENRY THOMPSON, F.R.C.S., Emeritus Professor of Clinical Surgery and Consulting Surgeon to University College Hospital. Eighth Edition. 8vo, with 121 Engravings, 10s. 6d.

By the same Author.

Diseases of the Prostate:

Their Pathology and Treatment. Sixth Edition. 8vo, with 39 Engravings, 6s.

Also.

Surgery of the Urinary Organs.

Some Important Points connected therewith. Lectures delivered in the R.C.S. 8vo, with 44 Engravings. Students' Edition, 2s. 6d.

Also.

Practical Lithotomy and Lithotrity;

or, An Inquiry into the Best Modes of Removing Stone from the Bladder. Third Edition. 8vo, with 87 Engravings, 10s.

Also.

The Preventive Treatment of Calculous Disease,

and the Use of Solvent Remedies. Third Edition. Crown 8vo, 2s. 6d.

Also.

Tumours of the Bladder:

Their Nature, Symptoms, and Surgical Treatment. 8vo, with numerous Illustrations, 5s.

Also.

Stricture of the Urethra, and Urinary Fistulæ:

their Pathology and Treatment. Fourth Edition. With 74 Engravings. 8vo, 6s.

Also.

The Suprapubic Operation of Opening the Bladder for the Stone and for Tumours.

8vo, with 14 Engravings, 3s. 6d.

Electric Illumination of the Male Bladder and Urethra,

as a Means of Diagnosis of Obscure Vesico-Urethral Diseases. By E. HURRY FENWICK, F.R.C.S., Assistant Surgeon to the London Hospital and Surgeon to St. Peter's Hospital for Stone. 8vo, with 30 Engravings, 4s. 6d.

Modern Treatment of Stone in the Bladder by Litholopaxy.

By P. J. FREYER, M.A., M.D., M.Ch., Bengal Medical Service. 8vo, with Engravings 5s.

The Surgical Diseases of the Genito-Urinary Organs, including Syphilis.

By E. L. KEYES, M.D., Professor of Genito-Urinary Surgery, Syphiology, and Dermatology in Bellevue Hospital Medical College, New York (a revision of VAN BUREN and KEYES' Text-book). Roy. 8vo, with 114 Engravings, 21s.

The Surgery of the Rectum.

By HENRY SMITH, Emeritus Professor of Surgery in King's College, Consulting Surgeon to the Hospital. Fifth Edition. 8vo, 6s.

Diseases of the Rectum and Anus.

By W. HARRISON CRIPPS, F.R.C.S., Assistant Surgeon to St. Bartholomew's Hospital, &c. 8vo, with 13 Lithographic Plates and numerous Wood Engravings, 12s. 6d.

Urinary and Renal Derangements and Calculous Disorders.

By LIONEL S. BEALE, F.R.C.P., F.R.S., Physician to King's College Hospital. 8vo, 5s.

The Diagnosis and Treatment of Diseases of the Rectum.

By WILLIAM ALLINGHAM, F.R.C.S., Surgeon to St. Mark's Hospital for Fistula. Fifth Edition. By HERBERT WM. ALLINGHAM, F.R.C.S., Surgeon to the Great Northern Central Hospital, Demonstrator of Anatomy at St. George's Hospital. 8vo, with 53 Engravings. 10s. 6d.

Syphilis and Pseudo-Syphilis.

By ALFRED COOPER, F.R.C.S., Surgeon to the Lock Hospital, to St. Mark's and the West London Hospitals. 8vo, 10s. 6d.

Diagnosis and Treatment of Syphilis.

By TOM ROBINSON, M.D., Physician to St. John's Hospital for Diseases of the Skin. Crown 8vo, 3s. 6d.

By the same Author.

Eczema: its Etiology, Pathology, and Treatment.

Crown 8vo, 3s. 6d.

Coulson on Diseases of the Bladder and Prostate Gland.

Sixth Edition. By WALTER J. COULSON, Surgeon to the Lock Hospital and to St. Peter's Hospital for Stone. 8vo, 16s.

The Medical Adviser in Life Assurance.

By Sir E. H. SIEVEKING, M.D., F.R.C.P. Second Edition. Crown 8vo, 6s.

A Medical Vocabulary:

An Explanation of all Terms and Phrases used in the various Departments of Medical Science and Practice, their Derivation, Meaning, Application, and Pronunciation. By R. G. MAYNE, M.D., LL.D. Sixth Edition, by W. W. WAGSTAFFE, B.A., F.R.C.S. Crown 8vo, 10s. 6d.

A Dictionary of Medical Science:

Containing a concise Explanation of the various Subjects and Terms of Medicine, &c. By ROBLEY DUNGLISON, M.D., LL.D. Royal 8vo, 28s.

INDEX.

[Continued on the next page.

The following CATALOGUES issued by J. & A. CHURCHILL will be forwarded post free on application :—

A. *J. & A. Churchill's General List of about* 650 *works on Anatomy, Physiology, Hygiene, Midwifery, Materia Medica, Medicine, Surgery, Chemistry, Botany, &c., &c., with a complete Index to their Subjects, for easy reference.* N.B.—*This List includes* B, C, & D.

B. *Selection from J. & A. Churchill's General List, comprising all recent Works published by them on the Art and Science of Medicine.*

C. *J. & A. Churchill's Catalogue of Text Books specially arranged for Students.*

D. *A selected and descriptive List of J. & A. Churchill's Works on Chemistry, Materia Medica, Pharmacy, Botany, Photography, Zoology, the Microscope, and other branches of Science.*

E. *The Medical Intelligencer being a List of New Works and New Editions published by J. & A. Churchill.*

[Sent yearly to every Medical Practitioner in the United Kingdom whose name and address can be ascertained. A large number are also sent to the United States of America, Continental Europe, India, and the Colonies.]

AMERICA.—*J. & A. Churchill being in constant communication with various publishing houses in Boston, New York, and Philadelphia, are able, notwithstanding the absence of international copyright, to conduct negotiations favourable to English Authors.*

LONDON: 11, NEW BURLINGTON STREET.

Pardon & Sons, Printers,]   [*Wine Office Court, Fleet Street, E.C.*

www.ingramcontent.com/pod-product-compliance
Lightning Source LLC
Chambersburg PA
CBHW022002190326
41519CB00010B/1359